RAISING YOUR OWN TURKEYS

Leonard S. Mercia

A Storey Publishing Book

The mission of Storey Communications is to serve our customers by publishing practical information that encourages personal independence in harmony with the environment.

Photographs by Erik Borg
Illustrations by Cathy Baker

Printed in the United States by Capital City Press
20

Library of Congress Cataloging in Publication Data

Mercia, Leonard S.
Raising your own turkeys.

1. Turkeys. I. Title.
SF507.M47 636.5'92 81-6353
ISBN 0-88266-253-8 (pbk.) AACR2

Contents

CHAPTER 1

Getting Started

Fresh turkey.

Having fresh turkey on your kitchen table is only one reason you'll enjoy raising a small flock of these curious birds. Raising turkeys can also be an excellent family or youth project, an excellent way to develop an understanding of live animal management. Then, too, you may be able to build a profitable market for at least a small number of turkeys.

Turkeys are interesting birds to raise. Members of 4-H clubs seem to enjoy the turkey project as much as, or more than, any other poultry project. They have good results with them, too. Turkeys are fun to raise because they are very friendly and, given the opportunity, will follow you and observe your every move. Not only are they curious, at times, they're protective. When strangers approach, they alert you in their noisy, gobbly fashion.

Turkeys are not as difficult to raise as many think. They do require a little special care to get them off to a good start. Sometimes they are a little slow in learning to eat and drink. Turkeys should be isolated from chickens and other poultry to prevent such diseases as Blackhead or Sinusitis and, of course, it's important that they be kept warm and dry during the first few weeks or brooding period. If you start with good stock and provide good feed, housing and management, you can raise turkeys successfully.

Before launching into production, though, even on a small scale, be aware of the costs. Day-old turkey poults are quite expensive and they consume a considerable amount of feed; thus, the cost of pro-

Large white turkeys, almost ready for the table.

ducing full-grown market turkeys is quite high. Table 1 presents the costs of raising a heavy roaster turkey.

Before starting a flock, check into the local laws and ordinances because zoning regulations in some areas prohibit keeping poultry of any kind. If you live close to neighbors, keep in mind that there are noises, odors and possibly fly problems associated with raising turkeys.

TABLE I
ESTIMATED PER-BIRD COSTS OF RAISING HEAVY ROASTER TURKEYS

Item		*Cost*
Poult (day old)		$1.50- 2.00
*Feed (75 lbs.)		7.50- 9.00
Brooding, electricity, litter, misc.		0.30- 0.40
	TOTAL	$9.30-11.40

Assumptions: Small flock with relatively high feed costs; poult costs are usually high; all feed purchased; no labor, housing equipment or interest costs included. (Costs are 1981 figures.)

*Feed consumption will depend upon variety, strain and age at processing.

VARIETIES

Several domestic varieties have been developed from the wild turkey. Those that are most important commercially are the large white (sometimes referred to as the Broad-Breasted Large White), the Broad-Breasted Bronze and the Beltsville Small White (sometimes called the Beltsville White). There are several other varieties available to those interested in other than the economic traits such as fast growth, conformation, efficient feed consumption, white feathers and body size. These varieties include White Holland, Black, Royal Palm, Bourbon Red and Narragansett.

Until recently, the Broad-Breasted Bronze (Figure 1-1) was the most popular variety. The bronze has a good growth rate, conformation or meatiness, feed conversion and most qualities demanded by the turkey industry. However, it also has dark pin feathers that detract from the dressed appearance. This is a distinct disadvantage that has led to the gradual replacement of the Bronze by white birds.

Within the fast-growing, heavy-roaster turkeys, hens usually reach

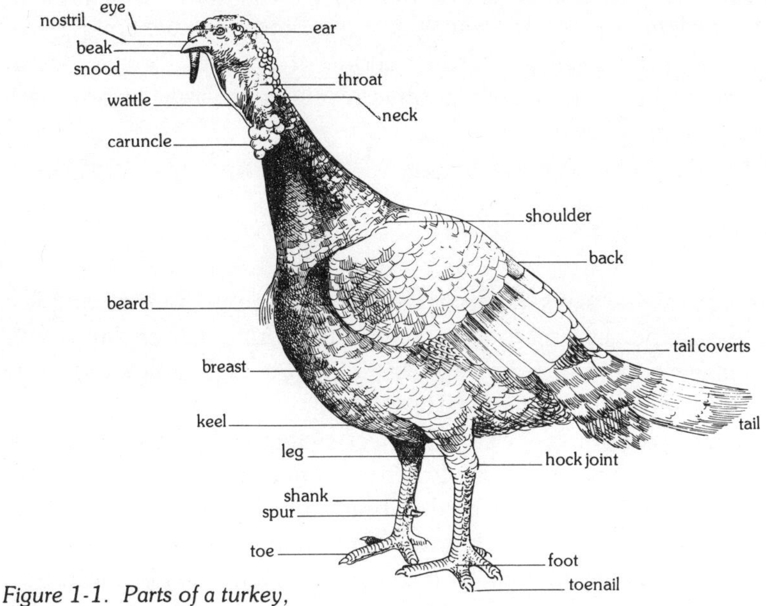

Figure 1-1. Parts of a turkey, shown here on a Bronze.

Large white turkey. Snood, wattle, caruncle, and ear are plainly visible.

a live weight of about 15 pounds at 18 weeks of age and toms weigh approximately 25 pounds at 20 weeks. The Beltsville White reaches good market condition in approximately 16 weeks and makes an excellent turkey broiler. At 16 weeks of age, the hens weigh approximately 10 pounds and the toms about 14 pounds. Smaller, fryer roasters can be produced by slaughtering either the small or large varieties at an earlier age.

BUYING POULTS

There are several ways to get started with a few poults but the simplest and least expensive method is to buy the poults, either day old or "started." Keeping breeders and hatching the eggs is an expensive, time-consuming method of starting a flock.

When buying poults, be sure to select a strain within the variety that is known to yield good results. Poultry strains are usually named after the breeder who has developed or bred one of the varieties such as Large White or Bronze. Each strain differs in those economic traits we look for such as growth rate, feed efficiency, low mortality, conformation and many others.

The new poults should originate from sources that are U.S. Pullorum-Typhoid clean and preferably from breeder flocks having no history of Sinusitis or air sac infection. Consult a state poultry specialist, a county Agricultural Extension agent, a commercial producer or some other knowledgeable person for advice on sources of good turkey poults in your area. It's hard to beat the flavor of a well-finished, mature turkey, especially one that is fresh. However, you may grow tired of turkey in time. The point to bear in mind is: don't raise more birds than you can eat and enjoy unless the surplus can be marketed profitably. In some cases, well-finished, fresh-dressed turkeys can be sold at prices substantially above their production costs.

Place your order for poults well in advance of the delivery date so as to be sure to get the stock you want. It is sometimes difficult to obtain small lots of turkeys delivered from the hatchery. However, it may be possible to pick the turkeys up at the hatchery or, perhaps, have a nearby producer order a few extra poults for you. Frequently, day-old or started poults can be purchased from local feed and farm supply outlets or specialized dealers who handle small lots of chicks, poults and other types of birds.

Although more expensive, started poults are often strong and hardy, so early poult losses are avoided. If six- to eight- week-old poults are purchased, they should no longer need brooding if the weather is warm. Sometimes poults shipped over long distances are subject to chilling or overheating. Buy the poults as close to home as possible to avoid these problems.

The best time to start the small turkey flock is in late May or June. Starting poults at that time will enable you to grow them to the desired market weights just prior to the traditional holiday season when the demand for turkey is strongest.

HOUSING

Because the small turkey flock is usually started in the warm months of the year, housing doesn't have to be fancy. However, the brooder house should be a reasonably well-constructed building that can be readily ventilated (Figures 1-2 and 1-3). If a small building is not available, perhaps a pen space within a larger building can be provided.

The area should have good floors that can be easily cleaned and

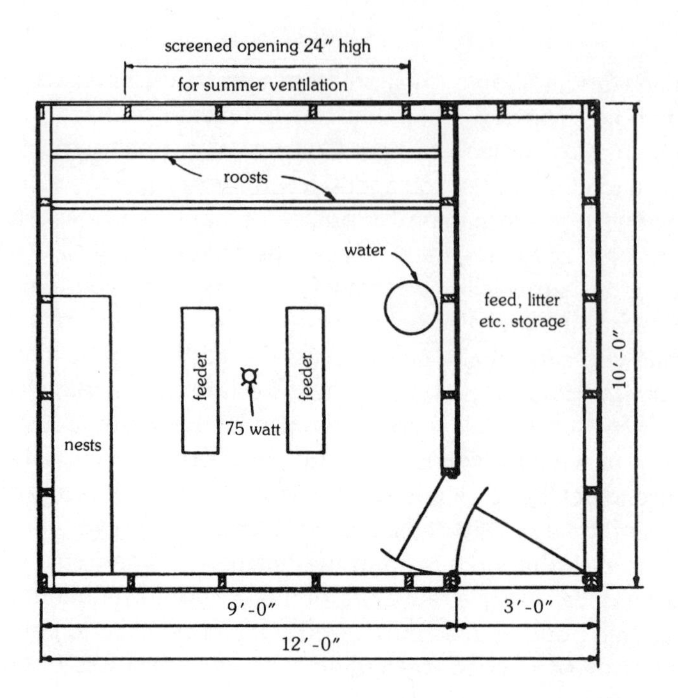

FLOOR PLAN

NOTE: Consult local health and building code authorities before starting construction.

BILL OF MATERIALS

Item	Material	Item	Material
Foundation	12 Conc. blocks 8″ × 8″ × 16″	Roof	175 bd. ft. T. & G. sheathing
Floor joists	7-pc. 2 × 6 × 10′		1½ sqs. roll roofing
Fr & rear sills	2-pc. 2 × 6 × 12′	Siding & doors	11 sheets 4″ × 8″ × ½″ ext. plywood
Floor	150 Bd. ft. T. & G. sheathing	Windows	2-pc. 2′ × 10′ fiberglass (flat)
Shoe	54 lin. ft. 2 × 4	Miscl. framing	4-pc. 2 × 4 × 12′
Studs: Rear	9-pc. 2 × 4 × 5′	Nails & hardware	
Front	9-pc. 2 × 4 × 7′	Rafters	7-pc. 2 × 6 × 12′
Ends	8-pc. 2 × 4 × 12′	Anchors-steel	6-pc. 1½″ × ¼″ × 12″
Partitions	2-pc. 2 × 4 × 12′	Fascia	2-pc. 1 × 6 × 12′
Plates	2-pc. 2 × 4 × 12′	Door stops	2-pc. 1 × 2 × 12′
			1-pc. 1 × 2 × 6′

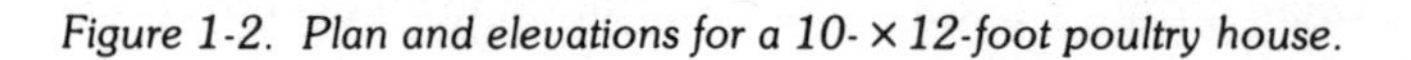

Figure 1-2. Plan and elevations for a 10- × 12-foot poultry house.

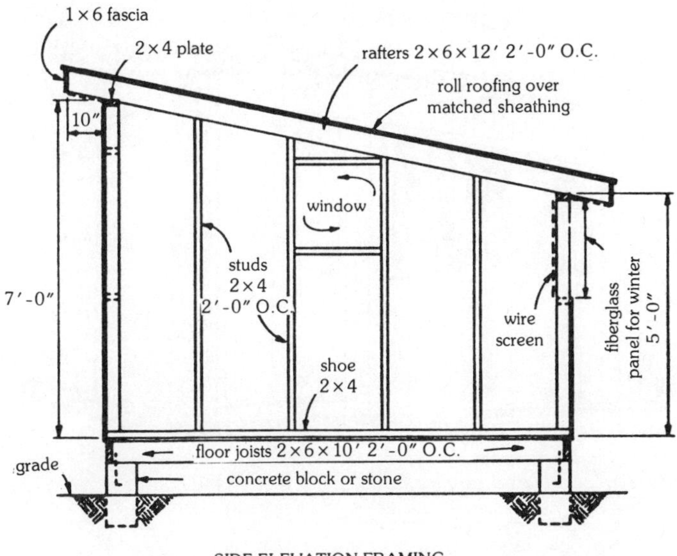

SIDE ELEVATION FRAMING

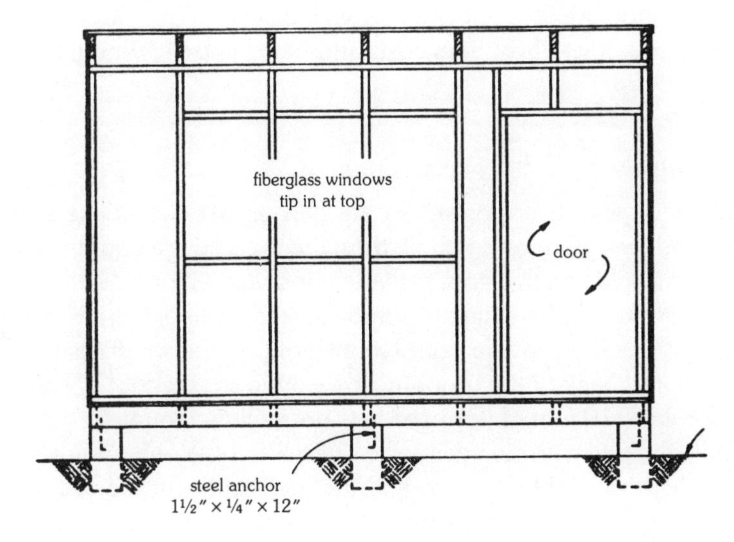

FRONT ELEVATION FRAMING

Figure 1-3. The completed poultry house from Figure 1-2.

disinfected. Cement floors are preferred, but wood floors are acceptable. The amount of insulation required in the building will depend upon the time of the year that the turkey poults are started, as well as climatic conditions in your area. A well-insulated building will conserve energy, lower brooding costs and help keep the young turkeys warm and dry.

Lighting

Adequate light is important so the pen or brooder house should have windows. Windows that tilt from the top and are equipped with anti-draft shields on the sides provide good ventilation. It's important that windows can be regulated and located to avoid drafts on the young poults. To provide cross-ventilation, locate windows in both the front and back of the brooding area. One square foot of window area for each 10 feet of floor space is normally adequate.

Equip the pen with electricity and artificial light. Young poults need intense light to enable them to find the feed and water, and thereby prevent starvation or dehydration. For the first two weeks, provide a minimum of 12 to 15 foot candles of light at the poult level. (One hundred and fifty watts of light with aluminum pie plates as reflectors located over the brooding area will provide adequate light intensity.) Bright light should be used 24 hours a day for the first three days. A dim night light is usually provided thereafter to prevent piling of the confined birds. Depending on environmental conditions, brooding of the poults is usually completed after five or six weeks. They are then ready for their permanent growing quarters.

It's important to provide adequate floor space for poults to avoid such problems as cannibalism. For heavy varieties, provide 1 square foot of floor space per poult up to eight weeks of age. From eight to twelve weeks, increase the floor space to 2 square feet per poult, and from 12 to 16 weeks, allow 2½ square feet minimum.

Mixed sexes grown in confinement need 4 square feet of floor space per bird from 16 weeks to market. If the flock is all toms, provide 4½ to 5 square feet of floor space; if all hens, 3 square feet is adequate. For light-type turkeys, the floor-space requirements may be reduced slightly.

Clean and disinfect the area to be used for brooding poults thoroughly. Many good disinfectants are available from agricultural supply houses. Whatever disinfectant is used, follow the directions on the container. Some materials can cause disinfectant injury to feet or eyes; this may damage the poults severely.

Management Systems

Several methods, or "management systems," are suitable for raising turkeys. *Confinement rearing*, in an enclosed poultry house, is frequently used by the small-flock grower. Where predators or adverse weather conditions are likely, or where there is limited range or yard area, confinement rearing is necessary. When possible, many small-flock owners provide a *yard* for their birds within a fenced-in area, using the brooder house as a shelter. Where a separate *range area* is available, it may be fenced in and range shelters with roost provided.

The management system of choice depends upon several factors including personal preference, the availability of adequate housing space or range area. It is safer to grow the poults in confinement. However, if building space is not adequate to grow birds to market age then other arrangements must be made.

For some, this will mean adding a *porch* to the brooder house. The porch has the advantage of keeping the birds out of their droppings which may cause disease problems. On the other hand, foot and leg problems may be more of a difficulty with the use of porches.

Yards can be used to good advantage to enable one to grow more birds in a given housing area. Keeping the yard clean may be a problem.

Range rearing requires good fencing and range shelters and feeders and waterers that can be moved frequently. With a good range some feed may be saved. However, thefts, losses from predators and other problems such as more labor may offset the advantages.

Sun porches were once very popular with turkey growers and are still used by some producers (see Figure 1-4). Large commercial producers tend to favor confinement rearing or range rearing rather than the use of porches. Sun porches are usually attached to the brooder house or shelter. The floor of the porch is made of either slats or wire. The porch is elevated to provide space underneath for the accumulation of droppings and an easy access for cleaning. The porches are fenced in on the top and sides. Coarse mesh wire is used for the top where snow loading may be a problem. Porches are sometimes used to help accustom the birds to changes in weather conditions or extremes of temperature prior to going on range. Another management system is the use of paved yards, or gravel or stone-surfaced yards adjacent to the brooder or growing house.

A small turkey porch with wire sides and floor.

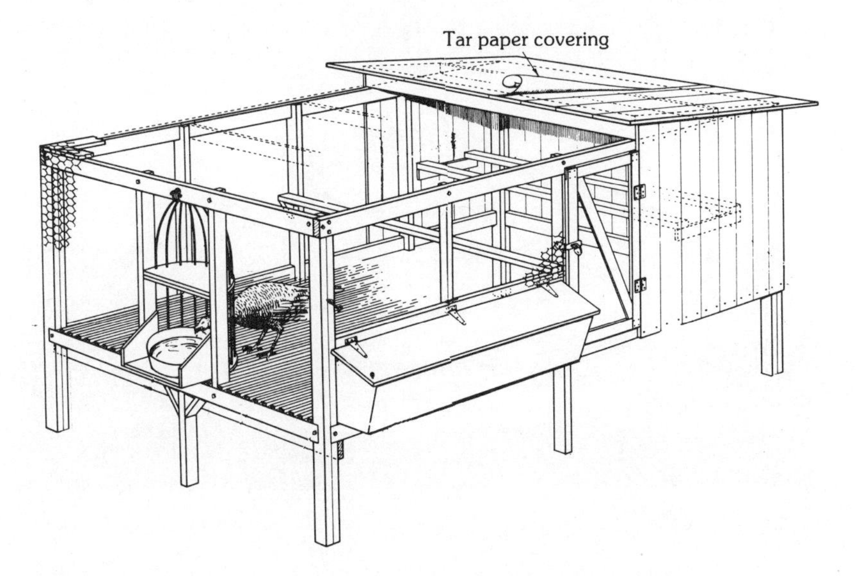

Figure 1-4. A home-made sunporch. For detailed plan, see Figure 3-6.

Large producers frequently use windowless, controlled-environment houses. Usually, these buildings are well insulated to achieve maximum efficiency in the use of heat during the brooding period and to provide comfortable, well-ventilated conditions for the birds. For more detailed information on housing and management systems, see Chapter 3.

EQUIPMENT

Brooding Equipment

Several types of brooders are suitable for brooding poults. The heat source may be gas, electric, oil or even wood or coal. Electric or gas brooders are probably best suited for the small flock. If a hover-type brooder, one with a canopy over the heat source, is used (Figure 1-5), allow approximately 12 to 13 square inches of hover or canopy space per poult. A brooder rated for 250 chicks is adequate for up to 125 turkey poults.

When employing hover-type brooders, use a small attraction light

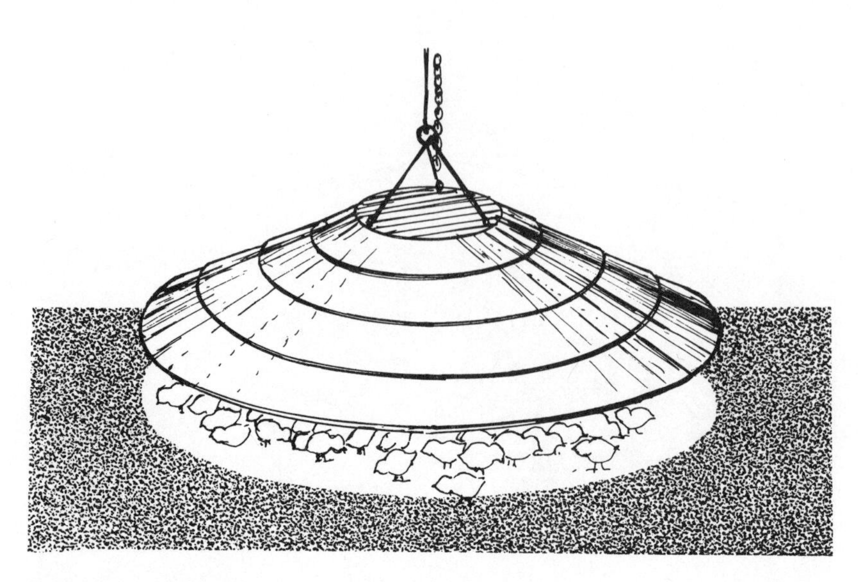

Figure 1-5. A commercial hover-type brooder.

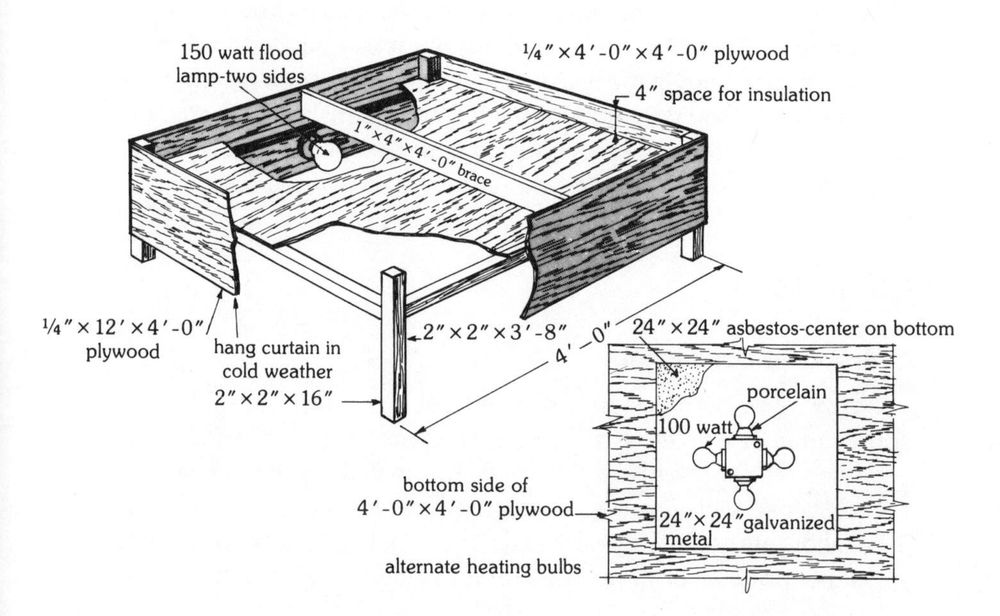

Figure 1-6. A home-made hover-type brooder.

Figure 1-7. An infrared brooder lamp.

under the hover to attract the poults to the heat source. The brooder should be equipped with a thermometer that can be read easily. Take temperature readings at the edge of the hover approximately 2 inches above the floor.* A home-made, hover-type brooder is shown in Figure 1-6. The brooder size may be adjusted to meet your needs. For example, an 18- × 18-inch brooder would be suitable for 25 birds.

Infrared brooders are very satisfactory for small numbers of poults (Figure 1-7). Provide two or three 250-watt bulbs per 100 poults. Even though one lamp may be adequate for the number of poults started, an additional bulb is recommended as a safety factor in case of a bulb failure. Hang infrared lamps about 18 inches from the surface of the litter at the start. After the first week, raise them 2 inches each week until they reach a height of 24 inches above the litter. The room temperature *outside* the hover or brooder area should be approximately 70° F. for maximum poult comfort.

* For a discussion of brooder temperatures, see Chapter 2.

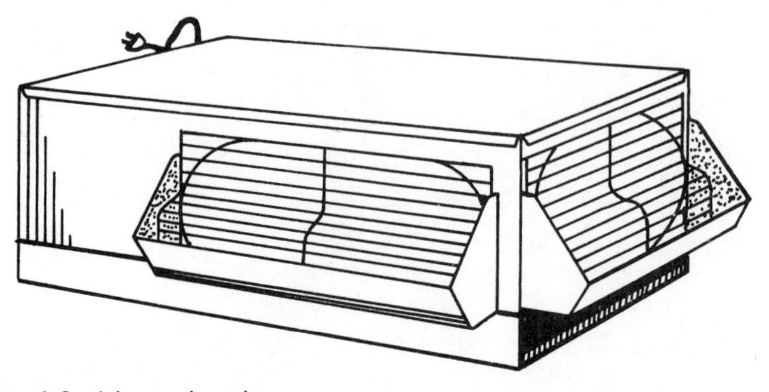

Figure 1-8. A battery brooder.

Where available, battery brooders can be used for the first seven to ten days to get the poults started. Allow approximately 25 square inches of space per poult in the battery. After the poults are removed from the battery and placed on the floor, watch them carefully to make sure they learn to use the feeders and waterers. A battery brooder is shown in Figure 1-8.

Feeding Equipment

To help get the poults eating quickly, place their first feed on egg filler flats, chick box lids, paper plates, small plastic trays, or box covers (Figure 1-9). When the poults arrive, place them in the feed container. When one or more starts pecking at the feed, the others will be attracted to it.

It is important to get poults started early on feed and water. They don't find the feed and water easily and starvation or dehydration can occur. When small, chick-sized feeders are used as the first feeders, bright-colored marbles, Christmas light bulbs, or other colored objects are sometimes placed in the feed and water containers. These help attract the poults to the feed and water. Oatmeal or fine granite grit, sprinkled very lightly over the feed once or twice a day for the first three days, may also help to get them eating.

When box tops or egg filler flats are used as the early feeders, place them next to the regular-type feeders. Usually, at about seven to ten days, the early feeders can be removed and the poults will use the regular feeders. Sometimes, paper is laid underneath the feeders for the first few days to prevent litter eating. If the litter is not covered with paper be sure *not* to fill the feeder so full that it will overflow into the litter. This may lead to the practice of litter eating. The use of smooth-surfaced paper is not recommended since slippery surfaces, used over prolonged periods, can cause foot and leg problems in young poults. If available, use paper with a rough surface.

From seven days to three weeks of age, use small feeders. Provide 2 linear inches of feeder space per bird. From three weeks to market, the poults should have access to larger feeders about 4 inches deep and providing 3 linear inches of feeder space per bird. Hanging tube-type feeders are excellent for turkey poults. The amount of tube-type feeder space can be determined by multiplying the diameter of the feeder pan by 3⅐ to obtain the number of inches of feeder space available. When figuring feeder space, remember to multiply the hopper length by two if the poults are able to use both sides of the feed hopper. Thus, a 4-foot trough feeder actually provides 8 linear feet of feeder space. Various types of feeders which may be purchased or built at home are shown in Figure 1-10.

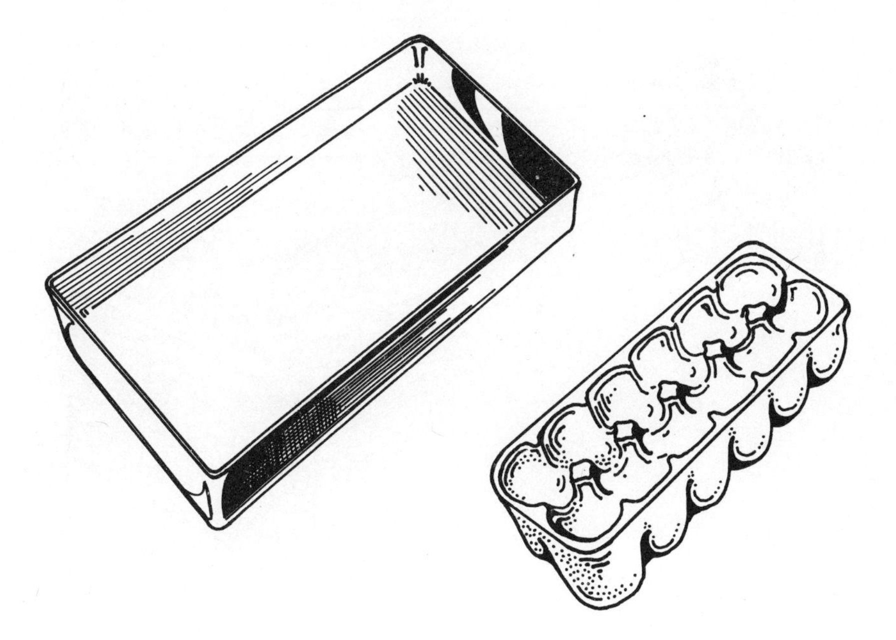

Figure 1-9. Shallow pans or egg cartons are fine for beginning feeders.

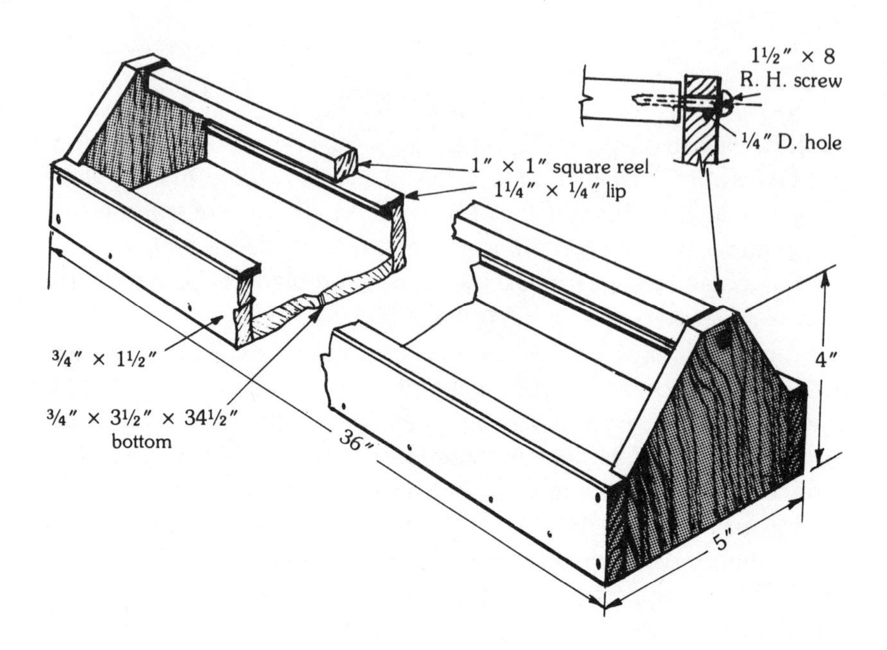

Figure 1-10. Various types of purchased and home-made feeders.

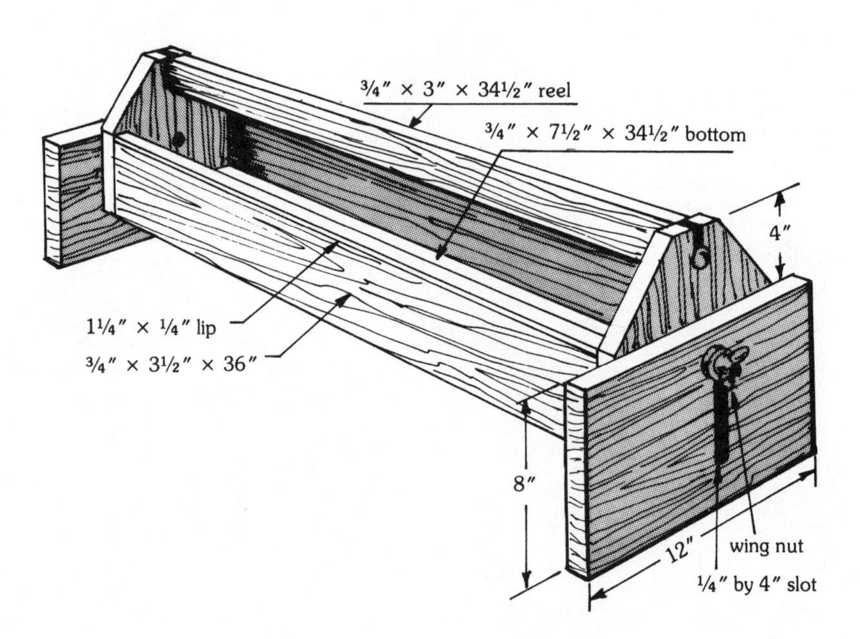

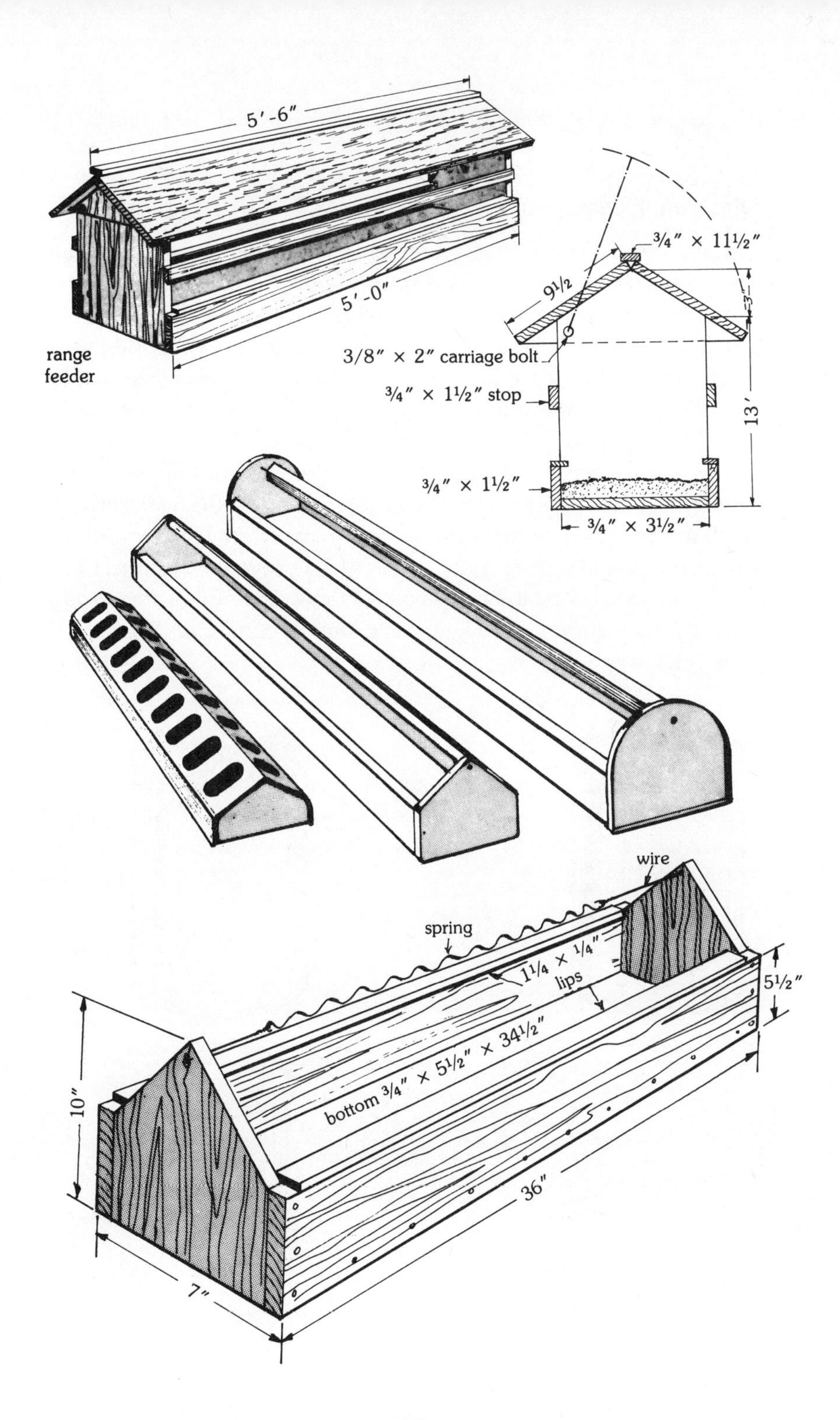
5′-6″
5′-0″
range
feeder
3/4″ × 11½″
9½
3″
3/8″ × 2″ carriage bolt
3/4″ × 1½″ stop
13′
3/4″ × 1½″
3/4″ × 3½″
wire
spring
1¼ × ¼″
lips
5½″
10″
bottom 3/4″ × 5½″ × 34½″
36″
7″

Watering Equipment

Usually, poults are started on either glass or plastic fountain-type waterers or automatic waterers. From one day to three weeks, they should have access to three 1- or 2-gallon fountains per 100 poults. From three weeks to market they should have two 5-gallon fountains per 100 poults, one 4-foot automatic waterer, or its equivalent. For smaller flocks, adjust number and size of waterers as necessary (Figures 1-11 and 1-12).

Change equipment, both feeders and waterers, gradually so as not to discourage feed and water consumption. Place the waterers on wire platforms. (The dimensions of the wire platform will depend on the size and type of waterer used.) This helps prevent litter from fouling the waterers, keeps the poults out of the wet litter that frequently surrounds the waterers, and keeps the litter in better condition. See Figure 1-13.

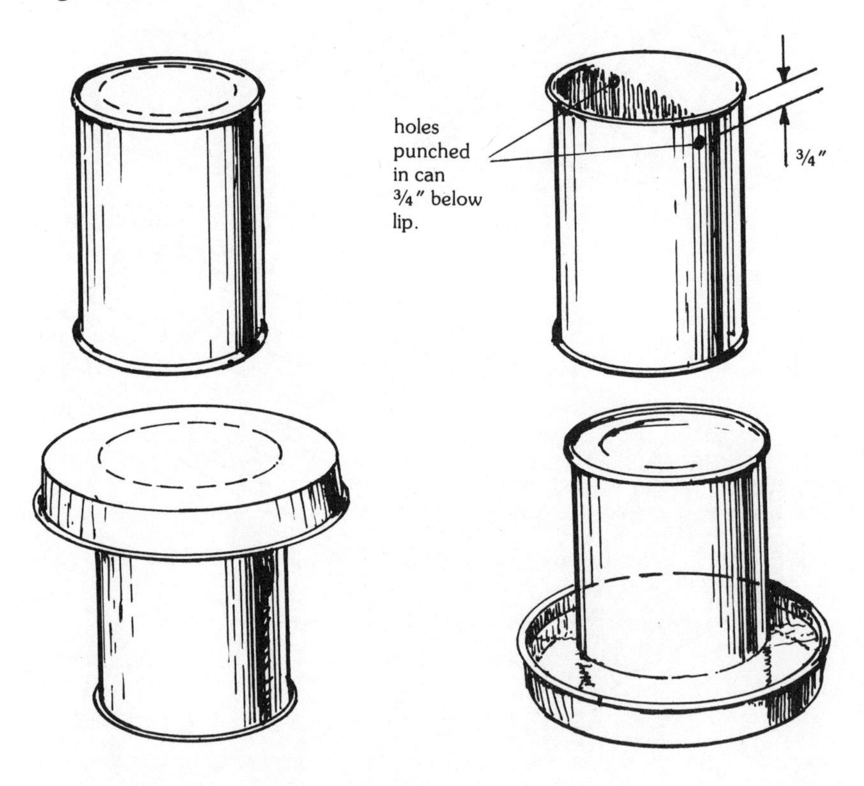

Figure 1-11. Home-made waterer made from a gallon oil can and a shallow pan.

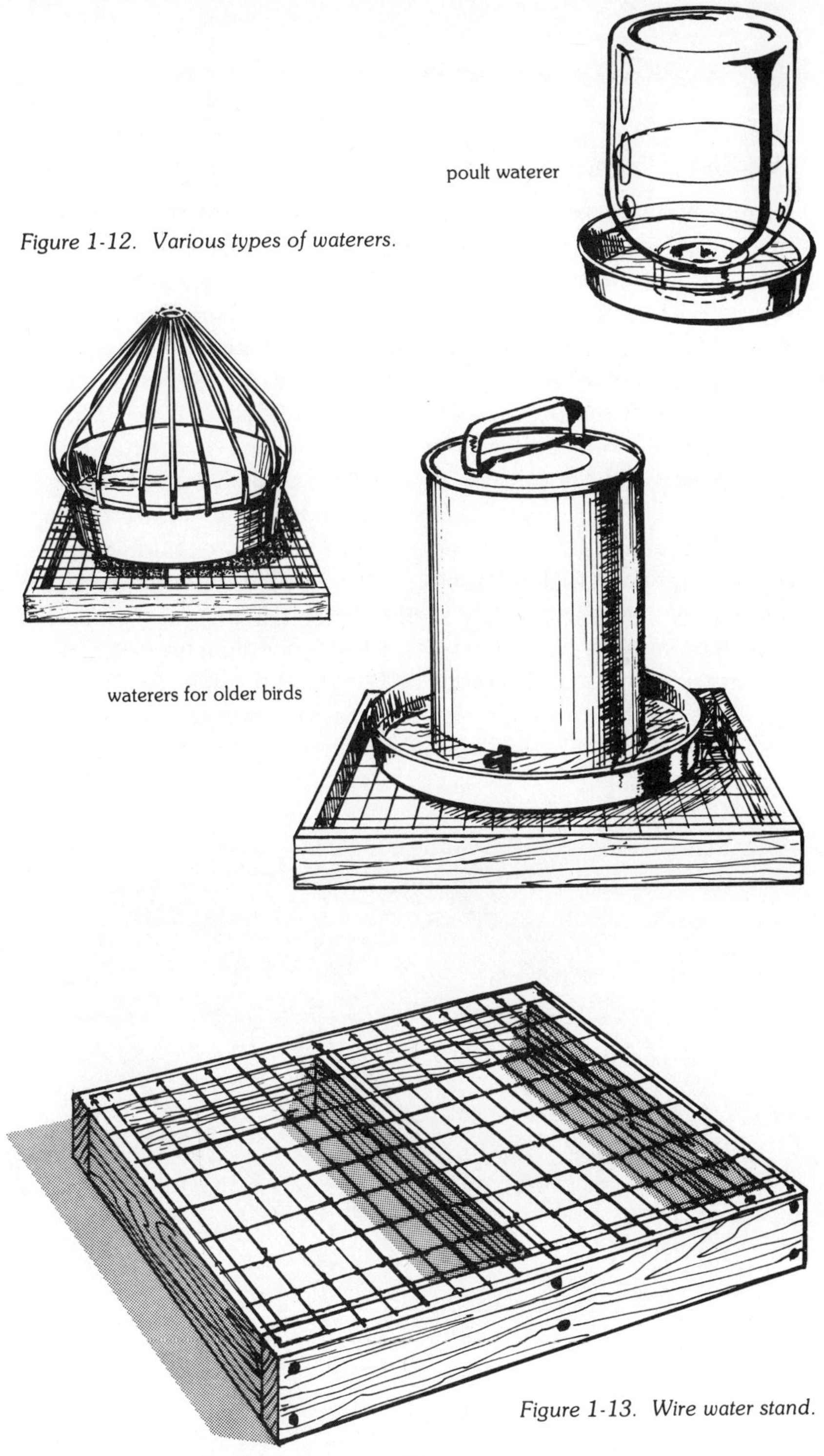

Figure 1-12. Various types of waterers.

Figure 1-13. Wire water stand.

Roosts

Roosts are not used frequently for turkeys during the brooding period, though they do help prevent piling at night. Sometimes flashing lights, sudden noises or rodents running across the floor can startle the poults. They may pile into a corner and cause injuries or smothering. If roosts are used, they may be either the step-ladder type or merely flat frames with perches on top of them. (See Figure 1-14.) Usually, the roosts or perches are made of round poles 2 inches in diameter or 2- × 2-inch or 2- × 3-inch material. They are located 12 to 15 inches above the floor when used in the brooder house. Each bird should have 6 linear inches of roost space by the end of the brooding period. Screen the sides and ends of roost pits to prevent the poults from gaining access to the droppings. Normally, the birds will begin to use roosts at about four or five weeks of age.

When turkeys are grown completely in confinement houses, roosts are normally not used. The birds bed down in the litter on the floor and, therefore, litter conditions have to be good to prevent problems such as breast blisters, soiled and matted feathers, or off-colored skin

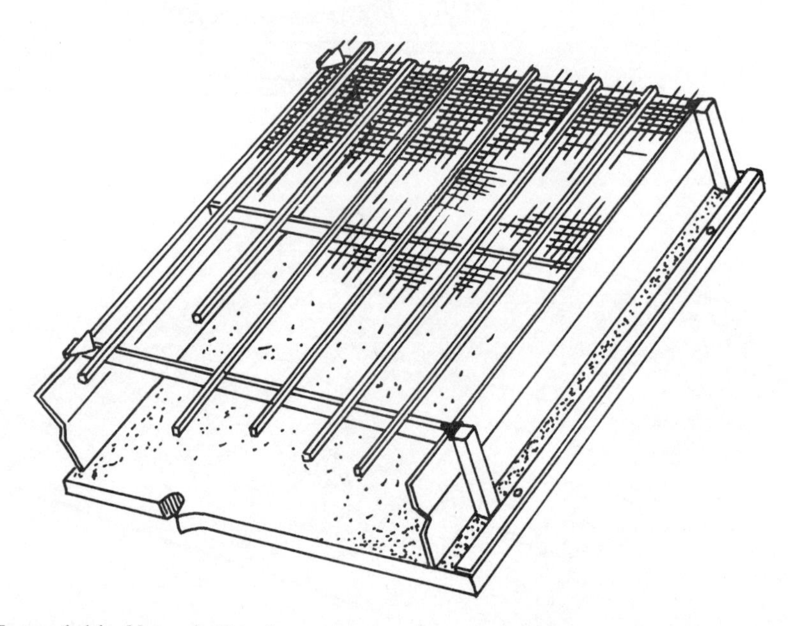

Figure 1-14. Home-built turkey roost.

blemishes on the breast. More importantly, good litter conditions also improve sanitation and prevent disease.

To maintain good litter conditions, remove wet or caked spots and replace with clean, dry litter material. Good pen ventilation will help remove excess moisture and keep the litter dry. Using good waterers located on stands will also help keep the litter dry.

Where birds are grown on range, roosts are used quite frequently and are constructed of 2-inch poles or 2- × 4-inch material laid flat with rounded edges. Space roost perches for range birds 24 inches apart and locate them 15 to 30 inches off the ground. If they are placed in a house or shelter, the roosts may be slanted to conserve space. Where an outdoor roosting rack is constructed, all roost perches may be built on the same level. Build this type of roost of fairly heavy material to prevent breaking when the weight of the birds is concentrated in a small area. Ten to 15 inches of perch space per bird are required for the large-type birds and 10 to 12 inches for the small-type turkeys up to maturity.

CHAPTER 2

Brooding the Poults

Brooding young turkey poults is one of the most enjoyable phases of turkey raising. During this period, the young poults grow rapidly beneath warm lights in a confined area. There are certain precautions you should take prior to placing the young poults in the brooding area.

PREPARING THE BROODER HOUSE

If the brooder house has been used for chickens or turkeys previously, it is very important to clean and disinfect the house and equipment before the young poults are put down. This means completely removing all litter and any caked material adhering to walls, floors or equipment. Wash the floors, sidewalls, ceilings and equipment thoroughly.

Then, disinfect the building and equipment using a disinfectant such as cresilic acid or one of the phenol or quaternary ammonium compounds. Disinfectants are available from farm-supply houses. Follow the label directions very closely to avoid disinfectant injury to the poults. After applying some of these materials, let the brooder house dry and air out for about two weeks prior to placing the poults in the house.

After the building is cleaned thoroughly, disinfected and dry, place about 3 to 4 inches of litter material on the floor. A good litter must be absorbent, light in weight, of medium particle size and a good insulator. Some of the common suitable litter materials are wood shavings, sawdust, sugarcane, ground corn cobs and peat moss. Litter material absorbs moisture from the fecal material as well as water spilled from drinking fountains. It also insulates the floor for bird comfort. Litter may be covered or uncovered. Some producers cover the litter with paper to prevent litter eating for the first week. As mentioned earlier, if the litter is covered, use a rough paper to prevent foot and leg problems. Distribute the litter very evenly over the floor and make sure it's dry and free of mold and dust. Very coarse litter material can also contribute to leg disorders, while fine materials can be too dusty.

It is usually a good idea to round the corners of the brooder house with small mesh wire, or a solid material to prevent piling in the corners. The brooder house should have a good roof to prevent wet litter. Plug holes in the building to prevent rats or any other animals from gaining entrance to the pen.

Brooder Guards

Use brooder guards or brooder rings to confine the birds to the heat source and the feeding and watering equipment until they're accustomed to their environment. The brooder guard also prevents drafts on the poults. The brooder guard should be 14 to 18 inches high; it can be made of corrugated cardboard, available in rolls several feet long. It may be tempered masonite or prefabricated panels that clip together to form a ring around the brooding area (Figure 2-1). For warm-weather brooding, or in houses where drafts are not a problem, the brooder guard can be made of poultry wire on frames.

Management of the brooder guard varies depending upon the design of the house and the climatic or seasonal conditions. When a non-insulated house is used during fairly cool weather (50°F. or below), an 18-inch brooder guard for each stove or heat source is recommended. For warm-weather brooding, a 12- or 14-inch brooder guard is satisfactory. Locate the brooder guard 2 to 3 feet from the edge of the heat source at the start and gradually move it out to a distance of 3 or 4 feet. Remove the guard the tenth day. Set up the brooder guard carefully so that at least 6 to 12 inches of space exists between feeders or waterers and the brooder guard; this permits traffic around the ends of the feeders. Figure 2-1 shows a typical brooder set-up.

Establish the brooder about 48 hours before the time the poults are

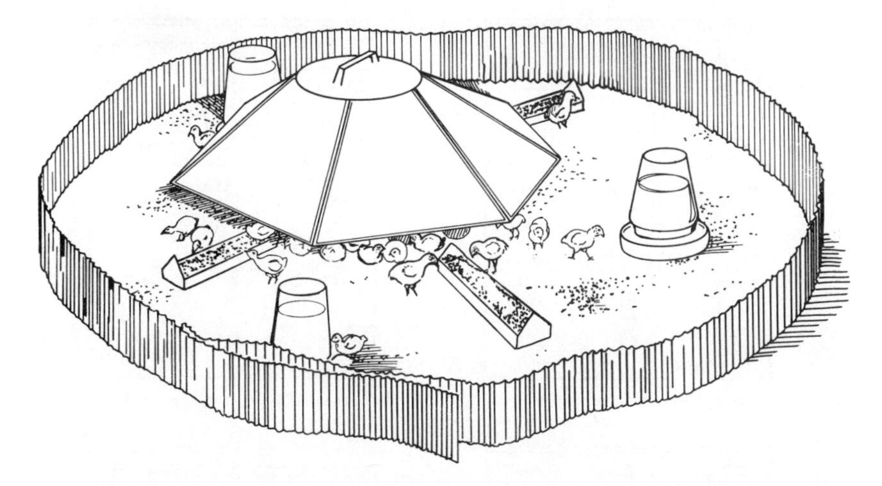

Figure 2-1. A brooder guard rings around the brooder area. Remember, young poults soon learn to hop and fly, so place your brooder in a confined area, well protected from cats, dogs and predators.

due to arrive. Fill the feeders and waterers and have everything ready so the poults can be removed from the containers as soon as they arrive. Again, it is important that they be put onto feed and water as quickly as possible. When a small number of poults is involved, remove each poult from the shipping box, and dip its beak first in water and then in the feed. This encourages the poults to eat and drink.

If hover-type brooders are used, the temperature should be approximately 95°F. the first week. Take this temperature reading at the edge of the hover approximately 2 inches above the litter, or at the height of the poult's back. Be sure to check the accuracy of the thermometer before the poults arrive. Reduce the hover temperature approximately 5° weekly until it registers 70° or 75°F. or is equivalent to the prevailing environmental temperature.

If the weather is warm during the brooding period, heat may be shut down during the day after the first week. Heat during the evening hours will be required for a longer period. Normally, little or no heat is required after the sixth week depending upon the time of year, the weather conditions and the housing. After the first week or so, experienced poultrymen can watch the poults and tell whether they are comfortable or not (See Figure 2-2). It may be necessary to turn on heat on cool evenings to keep the birds comfortable.

Brooder house drafty.

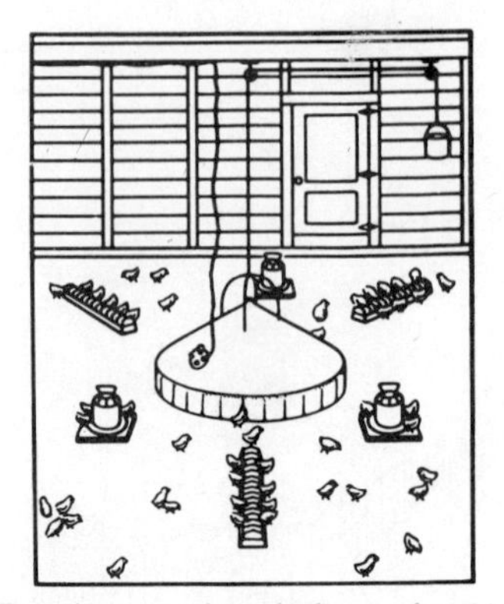

Brooder just right—chicks comfortable.

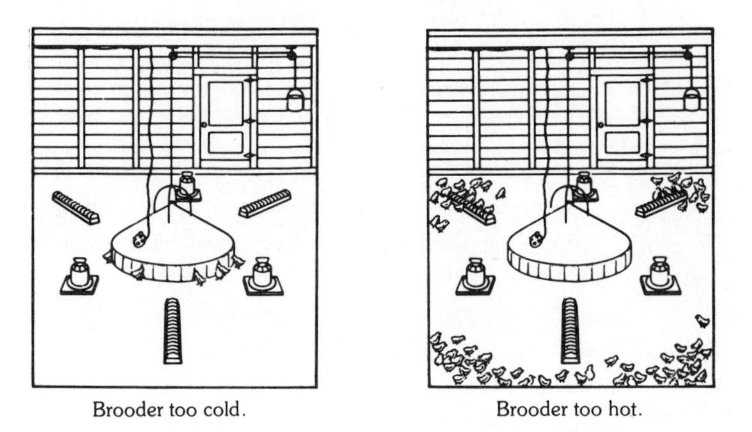

Brooder too cold. Brooder too hot.

Figure 2-2. Watch the poults to determine if the brooder is too cold or too hot.

Lighting

Provide high light intensity for the poults for the first two weeks of brooding to prevent starve-outs. If infrared brooders are used, they will provide adequate light intensity for the poults. If hover-type brooders are used, artificial light in addition to the attraction light should provide a minimum of 12 to 15 foot candles of light at the feeder and water level. Again, a reflectorized 150-watt light located over the brooding area will provide the necessary light intensity. Provide this light intensity for the first three weeks. After that time, about 1 foot candle of light is adequate, and this lower intensity will help reduce nervousness and flightiness in the flock.

Floor Space

Provide one square foot of floor space per poult up to eight weeks of age. From eight to twelve weeks increase the floor space allowance to 2 square feet per poult. From twelve to sixteen weeks, the minimum allowance is 2½ square feet. When mixed sexes are to be kept in confinement during the entire growing period, provide 4 square feet per bird. If the flock is all toms, 5 square feet of floor space is desirable; or, if all hens, 3 square feet is adequate. For light-type turkeys, the floor space requirements may be reduced. It's important to observe space requirements to avoid cannibalism and feather pulling and to make sure birds get adequate feed and water.

CHAPTER 3

Managing Turkeys Effectively

It is not difficult to raise turkeys assuming you start with good poults and feed and manage them well. Good management is an important factor—without it, optimal results will not be realized. Certain specific jobs and practices are peculiar to the turkey or to the management systems used. A management system is a method of raising and housing a flock.

Cannibalism

Feather picking and cannibalism are common problems in turkey flocks, especially when raised in close confinement. Proper flock management helps prevent these problems. Lack of floor space, lack of feed and water space, inadequate diets or a lack of feed or water can cause feather picking or cannibalism.

Probably the most effective means of controlling these vices is to debeak the birds. Debeak the poults when they are from three to five weeks of age. To delay beyond this period makes it difficult to handle the heavier birds and excessive feather picking may result. To prevent cannibalism and feather picking, debeaking should be a regular practice before noticeable picking occurs. Do not debeak the poults at day-of-age (the day after hatching) at the hatchery as this can interfere with the bird's ability to eat and drink and may cause starvation and dehydration.

The debeaking job is best done with an electric debeaker, a device with an electrically-heated blade that cauterizes as it cuts. (It may not be economically feasible for the small flock owner to own an electric debeaker. Frequently, these can be borrowed from a feed company or a neighboring poultryman.) When doing the operation for the first time, it may be advisable to get someone to demonstrate the method for you.

Insert your forefinger between the upper and lower beak. Depress the tongue under the finger to avoid burning it. For confined birds, remove about one half of the upper beak.

For birds going on range, debeaking is not usually done unless picking and fighting occurs. If it is necessary to debeak range birds, somewhat less is removed from the beak because excessive debeaking interferes with their grazing.

When a debeaking machine is not available, dog toenail clippers or heavy shears may be used. However, there is some danger of infection and bleeding when the beaks are not properly cauterized. Temporary control has been realized when the top beak is burned back

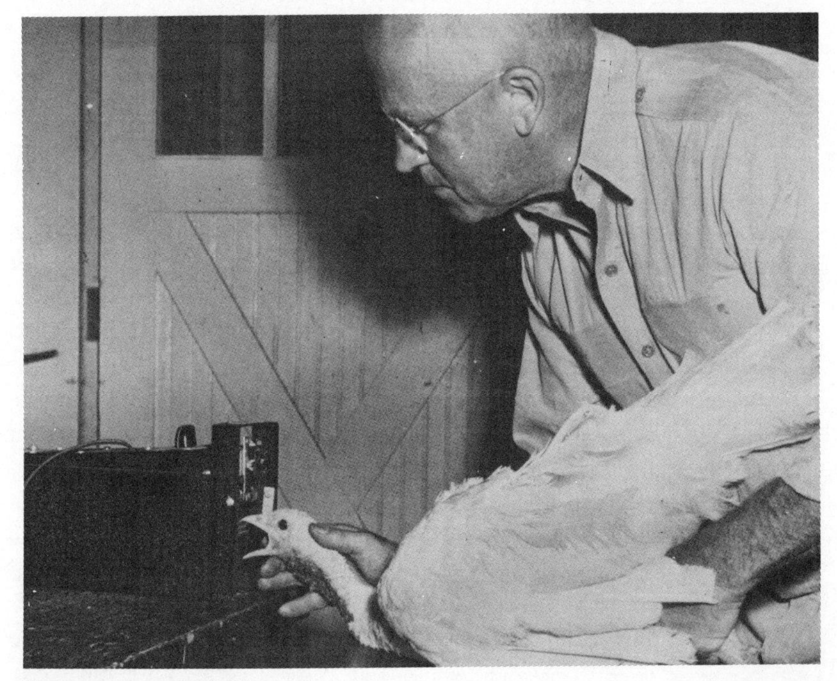

Using an electric debeaker.

Figure 3-1. A properly debeaked young turkey (left) and adult turkey (right).

slightly with a hot iron. When birds are debeaked, make sure feed and water levels are kept deep enough so the birds can consume adequate amounts of both. Figure 3-1 shows a properly debeaked bird.

Wing Clipping or Notching

Sometimes turkeys tend to fly, particularly when reared in the open or on range. When it is necessary to prevent the birds from flying, the wing feathers of one wing can be cut off with a sharp knife or scissors. Wing notching or the removal of the end segment of one wing with a debeaker is another method sometimes used to prevent flying. Wing notching can be done with an electric debeaker from one to ten days of age. Wing notching is not used as much now by commercial producers because, if improperly done, injuries can result when the birds attempt to fly. Injuries such as carcass bruises detract from the bird's dressed or oven-ready appearance.

Toe Clipping

Some turkey producers "toe clip" their birds to prevent scratches and tears of the skin on the bird's backs and hips that detract from the dressed appearance. Toe clipping is especially helpful in preventing scratches and skin tears when birds are reared in confinement under crowded conditions or are nervous. The practice also helps prevent carcass downgrading when birds are on range. The two inside toes are clipped so that the nails are completely removed. Surgical scissors or an electric debeaker may be used to remove the toes. It is advisable to have it done at the hatchery but you can do it yourself during the first few days.

General Management Recommendations

Keep young poults isolated from older turkeys and chickens. Take care to avoid tracking disease organisms from older stock to young stock or from other birds to the turkeys. Follow a good control program for mice and rats. These rodents are not only disease carriers but also consume large quantities of feed. Rats can kill young poults, too. Ideally, no other birds such as chickens, gamebirds or waterfowl should be on the same farm where turkeys are raised. However, with the medications available today it is possible to grow small flocks of turkeys where other types of birds have been kept, or are presently being kept. If abnormal losses or disease symptoms occur, take the birds immediately to a diagnostic laboratory for a positive diagnosis. (See Poultry Diagnostic Laboratories, p. 132.) One of the first symptoms of a disease problem is a reduction in feed and water consumption. So, watch daily feed and water consumption. If sudden changes occur that can't be traced to temperature or other stresses, look for disease problems.

With good management, you should be able to raise 95 percent of the turkeys started to maturity. If you're raising started poults (two weeks of age or older), the percentage should be somewhat higher. With high poult costs and feed costs, mortality can become very expensive, especially when the birds are lost during the latter part of the growing period.

Depending upon the disease exposure in the given area, it may be necessary to vaccinate the poults for such diseases as Newcastle, Fowlpox, Erysipelas or Fowl Cholera. To plan a vaccination program for your flock, check with the poultry diagnosticians at your state animal pathology laboratory, your county agricultural agent, your Extension poultry specialist, or some other knowledgeable person. Frequently, small flocks are not vaccinated and have no disease problems.

When used in the feed at preventive levels, medication effectively reduces losses from such diseases as Coccidiosis and Blackhead. An-

A good manager checks the turkeys frequently.

tibiotics and other drugs are of value in preventing and treating diseases, but don't use such medications as a substitute for good management. For more information on turkey diseases and their prevention, see Chapter 5.

MANAGEMENT SYSTEMS

As mentioned earlier, turkeys that have never been outside the brooder house may not seek shade from sunlight or shelter from the rain when placed on range. For this reason, producers who are going to range their turkeys frequently put their birds on sun porches attached to the brooder house. When the weather is warm, the young turkey poults can leave the brooder house and go out on the porches as early as three weeks of age. Usually, the porches are covered with fine mesh woven wire on the sides and top to prevent the poults from getting out. The floors may be either slat or wire. Either one works well if the birds will be on them for just a few weeks before going out on range. However, if birds are to be raised on porches up to market age, wire floors are not very satisfactory, particularly for heavier turkeys. Some birds tend to develop foot and leg problems as well as breast blisters or sores.

Smaller varieties of turkeys and those to be dressed at an early age for broilers or fryers do quite well on wire floors. Grow large varieties for heavy roasters on porches with slat floors.

Locating the feed and watering equipment so that they can be serviced from outside the porch greatly simplifies the chores. Many small turkey flocks are grown on porches very successfully. As with confinement rearing, birds grown on porches are not as likely to be attacked by predators as those grown on range or yard. They are also a lot less likely to develop disease problems, particularly the litter or soil-borne diseases. On the other hand, if adequate space is not available, and the birds are not debeaked, feather pulling and cannibalism tend to be more common with porch-reared birds.

Range Rearing

Range rearing offers an opportunity to reduce the cost of growing turkeys. This is especially true if the diet can be supplemented with home-grown grains. Turkeys are good foragers. And if good green feed is available on the range, this will mean less consumption of expensive mixed feed, thereby reducing the cost of the feeding program. Building costs are much less when birds are range-reared but labor requirements are greater.

Range rearing is not without its problems. Losses are possible from such things as soil-borne diseases, adverse weather conditions, predators and theft. Because of these potential problems and the additional labor required, confinement rearing has quite rapidly replaced range rearing in recent years.

Depending upon climatic conditions, some growers provide only roosts for turkeys on range. Some actually allow the turkeys to sleep on the ground. This management practice is more practical when the turkeys will be matured early and before the cold winter weather sets in. Portable range shelters give the turkeys much better protection during poor weather. They can be moved to new locations to provide the birds with better range conditions and prevent the development of muddy spots and contaminated areas. When portable shelters are used with roosting quarters, the feeders and waterers can be moved whenever the grass is closely grazed in an area.

Typical dimensions of portable shelters are 10 × 12 or 12 × 14 feet but they can be built smaller to accommodate small flocks. If built larger than this, they are not as easily moved and there is more chance that the building will be damaged when moved. Portable range shelters should provide a minimum of 2 square feet of space per large bird and 1.5 square feet for small-type birds. A 10 × 12-foot shelter can supply roosting space for up to 150 twelve-week-old turkeys and up to 75 mature birds. A typical range shelter is shown in Figure 3-2.

Normally, May- and June-hatched poults can be put out on range by eight weeks of age. Before putting them on range, make sure they are well feathered, especially over the hips and back. Check the weather forecast and try to move them out during good weather. It is best to move them in the morning to give them time to get adjusted to their new environment before darkness.

If possible, provide a range area that has been free of turkeys for at least one year, and preferably two. Poorly drained soil does not make a good range for turkeys. Stagnant surface water can be a source of disease. You can use a temporary fence to confine the flock to a small part of the range area. Move the fence once a week or as often as the range and weather conditions indicate. Permanent fencing may also

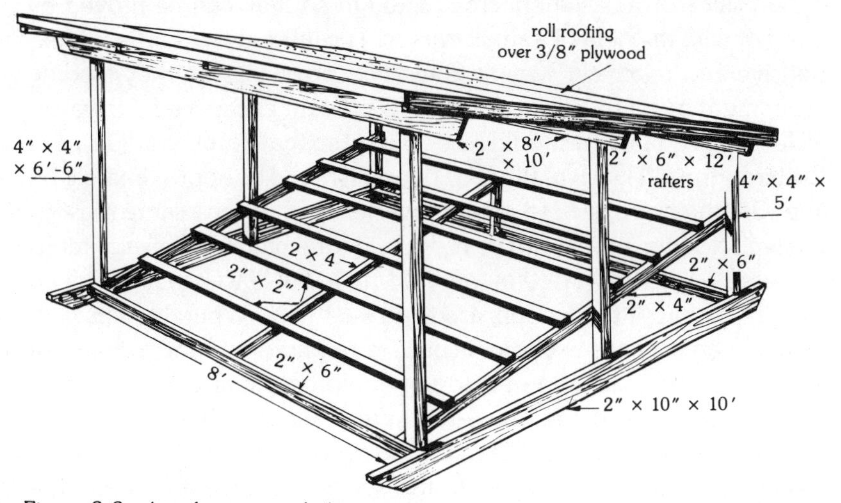

Figure 3-2. A turkey range shelter.

be necessary. Wild animals or dogs can cause losses on ranges by killing or injuring birds, or by causing stampedes resulting in suffocation and injuries. A 6-foot-high poultry fence around the range area will help prevent these problems. Some producers add electric fencing for more protection. A single wire, 6 inches from the ground outside the permanent fence, works well.

Provide artificial shade if there is no natural shade. Several rows of corn planted along the sunny side of the range area provide good shade and some feed. If range shelters are used, move them every seven to fourteen days, depending upon the weather and the quality of the range. Move the feed and watering equipment as needed to avoid muddy and bare spots.

The range crop selected will depend upon the climate, soil and the range management. Many turkey ranges are permanently seeded. Others are a part of a crop rotation plan. As part of a three- or four-year crop rotation, legume or grass pasture and annual range crops such as soybeans, rape, kale, sunflowers, reed canary grass, and sudan grass, have been used successfully. Sunflowers, reed canary grass and sudan grass provide green feed and also shade. For a permanent range, alfalfa, ladino clover, bluegrass and brome grass are very satisfactory.

Range feeders should be waterproof and windproof so that the feed is not spoiled or blown away. See Figures 3-3 and 3-4. Place the feeders on skids or make them small enough so they can be moved by hand or with the help of a small tractor. Trough-type feeders are inexpensive and relatively easy to construct. Specialized turkey feeding equipment can also be purchased. To prevent excessive feed waste, all feeding equipment should be designed so that it can be adjusted as the birds grow. The lip of the feed hopper should be approximately on line with the bird's back to prevent waste. And, for the same reason, the feed hopper should not be more than one half full. Pelleted feeds are less likely to be wasted on range. Provide at least 6 inches of feed trough per bird if the feeders are filled each day. When feeders with storage capacity are used, less space is required and the amount of feeder space should conform to the equipment manufacturer's recommendations. Provide one 4-foot automatic trough waterer or the equivalent per 100 birds. Clean the waterers daily and disinfect them weeky. Locate waterers close to the shelters. If possible, shade the

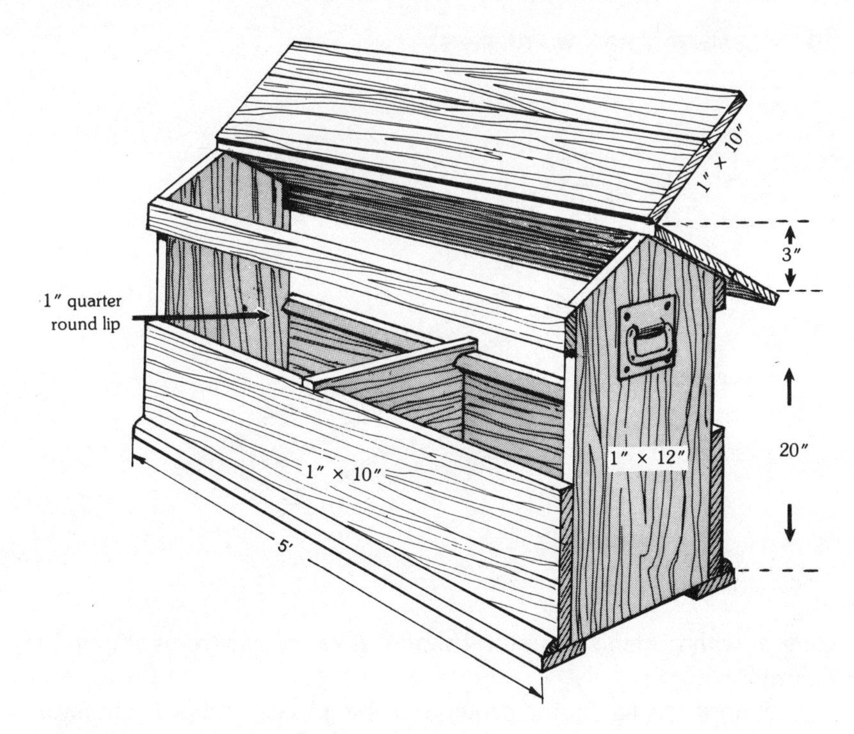

Figure 3-3. A small range feeder.

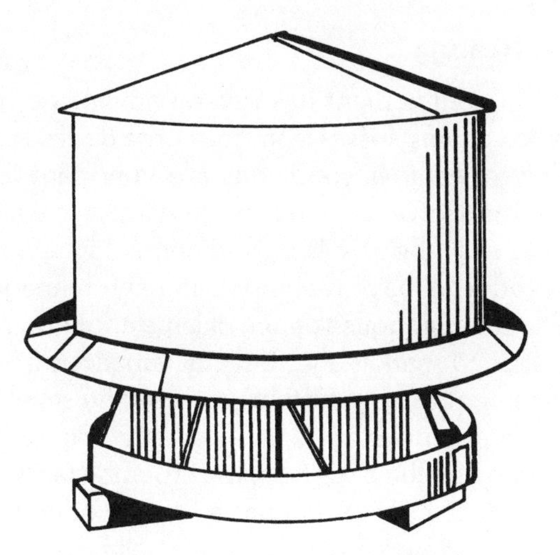

Figure 3-4. A hopper-type range feeder with rain guard.

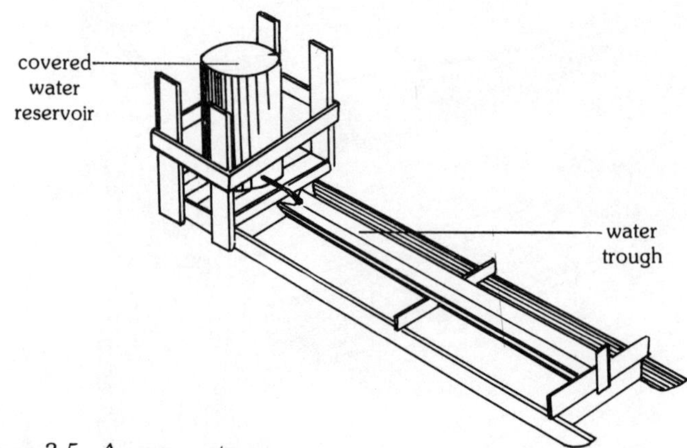

Figure 3-5. A range waterer.

waterer with portable or natural shade. A range waterer is shown in Figure 3-5.

The range shelter should provide roosting space and protection for the birds from wind, rain and sun. If natural shade is not available during hot weather, it is advisable to build a shelter to provide more shade.

Confinement Rearing

Rearing birds in confinement has several advantages over range rearing. Protection against losses from soil borne diseases, predators, thefts, and adverse weather conditions are important advantages. Lower labor costs and lower acreage requirements have also induced turkey growers to raise their birds in confinement in recent years.

Small flock producers have many housing and management-system options. The brooder house or pen, if large enough, may be used to confine the birds to market age. Several variations from the conventional confinement rearing systems are used by small flock owners; the house and porch system is one example (Figure 3-6). Some growers, after removing the birds from the brooding facilities, confine them to a wire-enclosed porch that has a roof for added protection. Ideally, a section of the porch should have a solid floor where a dry, fluffy litter can be maintained. This will help prevent the development of breast blisters or leg and foot problems.

When birds are raised in strict confinement, adequate floor space is important. If the birds are debeaked, if feed and water space are adequate, and other conditions are optimal, large males can be confined to approximately 5 square feet of floor space, females, 3 square feet and mixed flocks, 4 square feet. Smaller varieties need 4 square feet for males, 3 square feet for females and 3½ square feet for mixed flocks.

A large white turkey looks out from a sunporch.

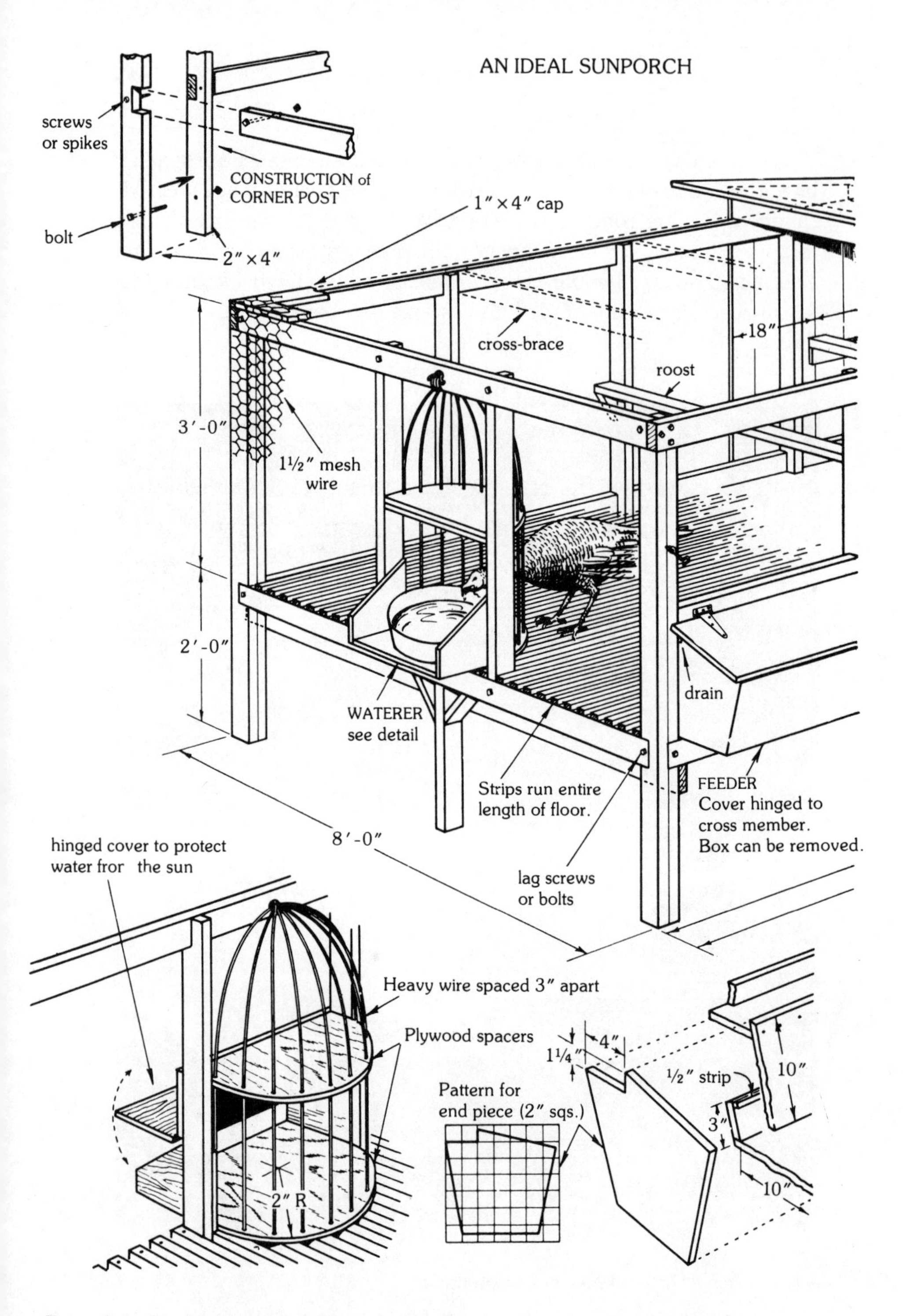

Figure 3-6. Plan and details for a house and sunporch system suitable for small flock.

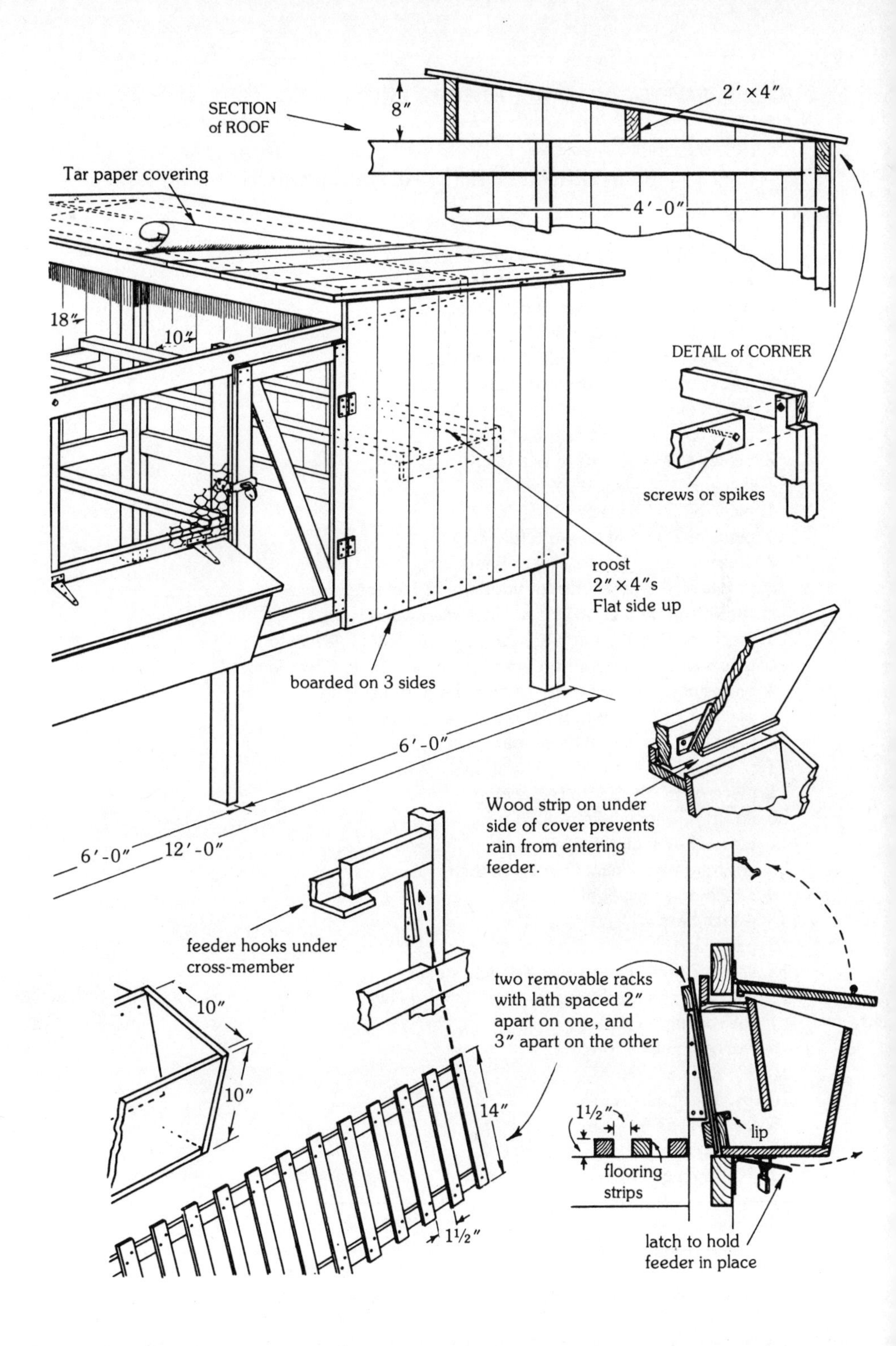
SECTION of ROOF
8″
2′×4″
4′-0″
Tar paper covering
18″
10″
DETAIL of CORNER
screws or spikes
roost
2″×4″s
Flat side up
boarded on 3 sides
6′-0″
6′-0″
12′-0″
Wood strip on under side of cover prevents rain from entering feeder.
feeder hooks under cross-member
two removable racks with lath spaced 2″ apart on one, and 3″ apart on the other
10″
10″
14″
1½″
1½″
flooring strips
lip
latch to hold feeder in place

MATERIAL FOR TURKEY SUNPORCH

5 pieces of 1 × 4 common board for door, 4 feet long
8 pieces of 2 × 4 corner posts 5 feet long
2 pieces of 2 × 4 center studs[1] 5 feet long
1 piece of 2 × 4 center stud 14 feet long
5 pieces of 2 × 4 floor joists[2] 8 feet long
2 pieces of 2 × 4 floor joists 12 feet long
2 pieces of 2 × 4 plates[3] 12 feet long
2 pieces of 2 × 4 plates 8 feet long
2 pieces of 2 × 4 roost 8 feet long
1 piece of 2 × 4 roost 12 feet long
1 piece of 2 × 8 rafter[4] 8 feet long
1 piece of 2 × 4 at feeder 6 feet long
2 pieces of 2 × 4 cross braces 8 feet long
33 pieces of 1½ × 1½ fir for flooring, 12 feet long
(or 96 square feet of 1½ × 1½ turkey wire mesh)
26 pieces of 1 × 8 common board, D4S or T & G, 4 feet long
14 pieces of 1 × 8 common board, D4S or T & G, 5 feet long
4 pieces of 1 × 12 common board, D4S, 6 feet long
1 piece of 1 × 12 common board, D4S, 8 feet long
2 pieces of 1 × 3 common board, D4S, 6 feet long
1 piece of 1 × 6 common board, D4S, 6 feet long
1 piece of ½ × ½ strip 6 feet long
50 lin. ft. of 2-inch lath for feed rack
1 piece of ¼-inch temp. prestwood, 3 feet × 4 feet
50 lin. ft. of heavy galv. wire, 8 or 9 ga.
3½-inch or 4-inch light T hinges
3 4-inch hasps[5]
3 padlocks
32 lin. ft. 2-inch poultry wire 48 inches high
12 lin. ft. 2-inch mesh poultry wire 36 inches high
1 roll roofing paper
5 lbs. 6d.[6] nails—com.
10 lbs. 8d. nails—com.
10 lbs. 10d. nails—com.
5 lbs. 16d. nails—com.
2 lbs. ¾-inch galvanized staples
1 lb. 1-inch galvanized roofing nails

---

[1]*stud*, upright
[2]*joist*, small beams laid horizontally to support floor
[3]*plate*, horizontal timber carrying rafters for roof.
[4]*rafter*, sloping timber of roof
[5]*hasp*, hinged metal strap secured by staples and pin
[6]*6d.* (6-penny size).

The Yard System

A variation of the range method of handling turkeys is the use of a yard that is attached to a brooder house, growing house or other type of building. Commercial producers sometimes provide pole buildings for shelter at night and let birds out on range during the day.

When birds are reared in confinement they need considerable floor space toward the end of the growing period. By using a yard attached to the housing facility, more birds can be kept in a smaller housing area. The yard should be well drained. Sometimes gravel or stones are put in the yard to keep the birds out of the mud, to improve sanitation and to prevent disease. Yards need to be kept in good condition. Ideally, the location of the yard should be changed every year or two. Four to five square feet of yard area per bird is recommended.

If there is a danger of predators, such as foxes or dogs, the range or yard should be fenced. Normally a woven-wire poultry fence 6 feet high will keep the turkeys inside, but in some cases it may be necessary to clip the flight feathers or primary feathers on one wing to prevent the birds from flying over the fence. Animals such as foxes, coons and dogs can cause considerable damage. Frequently, the harm's due more to piling and suffocation than outright deaths. Lighting the range with flood lights also helps keep animals out or discourages raiding of the range or yard by predators. Electric fencing also serves this purpose.

THE FEEDING PROGRAM

One of the best pieces of advice for turkey growers is that they select a good brand of feed and follow the manufacturer's recommendations for its use.

There are two basic feeding programs for turkeys: one is the all-mash system; and the second uses a protein supplement plus grains. The latter is frequently used in those areas where home-grown grains are available. Feed insoluble grit such as granite grit to the birds whenever grains are included as a part of the diet, or if the birds are on range. This enables them to grind and utilize the grains and other fibrous materials.

Feed-company recommendations for feeding turkeys vary considerably. However, they are all based on the fact that turkeys grow rapidly. They need a high-protein diet at the start to support this rapid growth. The nutrient requirements of turkey poults vary with age. As they become older, the protein, vitamin and mineral requirements decrease and the energy requirements increase.

One of the simpler feeding programs starts with a 28 percent protein-starter diet up to approximately eight weeks of age. The birds are changed to a 21 or 22 percent protein-growing diet and are fed this diet during the nine to sixteen week period. From 16 weeks to dressing time, they are fed a finishing diet containing approximately 16 percent protein (see Table 2). Other feed companies offer and recommend five or six different diets during the growing period. Again, buy feed from a reputable company and follow their recommendations. Some of the feeds will include medication such as a Coccidiostat or a Blackhead preventative. Observe the precautions on the feed tag and the recommended times for withdrawing the feed before dressing birds.

TABLE 2
FEEDING PROGRAM FOR GROWING TURKEYS

Age of Bird	*Type of Feed*	*Percent Protein*
0-8 weeks	Starter	27-28
9-16 weeks	Grower	20-21
16 weeks to market	Finisher	14-16

Based on New England College Conference Recommendations, revised 1980.

Keep feed and water before the birds continuously. When grains are fed as a part of the diet, the amount to be fed depends upon the protein content of the mash or pellets the birds are receiving. Remember that grains such as corn or oats contain approximately 9 to 10 percent protein. To feed the birds too much of these grains would dilute the total protein content of the combined diet to the extent that it could affect growth. During the 12 to 16 week period, the birds can receive a grain and mash diet, but don't dilute the protein content below the 16 percent level. When grains are fed with the finishing diet,

Filling the feeding trough with commercial pellets.

This trough will accommodate a dozen birds.

don't dilute the finishing diet with grains to the extent that it will reduce total protein intake below 14 percent.

Most turkey producers feed a nutritionally complete starter mash or pellet. However, you can use green feed for small flocks if labor requirements are not of concern. A tender alfalfa, white dutch clover, young tender grass or green grain sprouts, all chopped into short lengths and fed once or twice daily, are good for the poults. Turkeys like tender green feeds such as short, fresh lawn clippings or garden vegetables such as swiss chard, lettuce and even the outer leaves of cabbage. Turkeys should not be permitted to eat wilted, dry or long stringy roughage. These can cause impacted or pendulus crops. Again, if roughage materials are fed, make sure the birds receive an insoluble grit such as turkey-size granite grit.

Turkeys can be fed the concentrate in mash or pellet form. When changing from a mash to pellets, make the change gradually. A commercial concentrate may also be purchased and combined with ground grain or with soybean meal and ground corn in the proportion recommended by the manufacturer. Usually, the small grower will find it advantageous to use a complete, ready-mixed mash when the birds are reared in confinement. If the birds are on a good range, a complete feed, preferably in pellet form, supplemented with grains and insoluble grit, makes a sound feeding program. Growth-rate and feed-consumption information for heavy roaster turkeys is presented in Table 3. Water consumption information is presented in Table 4.

Feeder Management

Proper management of the feeders is important. Start the poults out on box tops, plates or small poult feeders. Provide larger feeders as the birds become older. If this is not done, the turkeys will "beak out" feed or knock the feeders over and feed is wasted in the litter. The same is true for waterers. As the birds grow, use larger waterers to avoid spillage and to make sure the poults get an adequate water supply.

One of the common problems seen by servicemen in the field is the waste of feed, and there are several reasons for it. Some of it can be due to sloppy handling, but by far the most important cause of feed loss or feed waste is the mismanagement of feeders. Never fill the feed hoppers more than half full, and, to avoid waste, keep the top or lip of

TABLE 3
GROWTH RATES AND FEED CONSUMPTION FOR RAPID-GROWING, HEAVY ROASTER TURKEYS

Age in Weeks	*Live Weight in Pounds*		*Total Cumulative Feed Required (Pounds)*		*Pounds of Feed per Pound of Live Weight*	
	Males	*Females*	*Males*	*Females*	*Males*	*Females*
2	0.6	0.5	0.7	0.6	1.2	1.2
4	1.8	1.5	2.5	2.1	1.4	1.4
6	4.0	3.5	6.3	5.6	1.6	1.6
8	6.8	5.7	11.4	10.3	1.7	1.8
10	10.0	7.9	17.6	15.5	1.8	2.0
12	13.1	10.2	25.7	22.0	2.0	2.2
14	16.2	12.2	36.5	28.7	2.3	2.4
16	19.2	13.8	47.0	35.7	2.4	2.6
18	22.4	15.2	60.5	43.9	2.7	2.9
20	25.5	16.6	74.2	52.1	2.9	3.1
22	28.6	17.6	86.9	59.5	3.0	3.4
24	31.6	18.6	102.1	69.2	3.2	3.7

SOURCE: *Poultry Management and Business Analysis Manual*, Bul. 566 (RU). Joint Publication, The New England Extension Poultry Specialists.

TABLE 4
DAILY WATER CONSUMPTION OF HEAVY ROASTER TURKEYS
(Gallons per 100 Birds)

Age in Weeks	*Water Consumption*	*Age in Weeks*	*Water Consumption*
1	1	11	14.0
2	2	12	15.0
3	3	13	16.0
4	4	14	16.5*
5	5	15	17.0*
6	6	16	16.5*
7	7.5	17	16.5*
8	9.5	18	16.5*
9	11.0	19	16.5*
10	12.5	20	16.5*

Dr. Salsbury Laboratories

*Consumption will vary from 15 weeks to maturity from 14 to 19 gallons per 100 birds per day depending upon environmental temperatures.

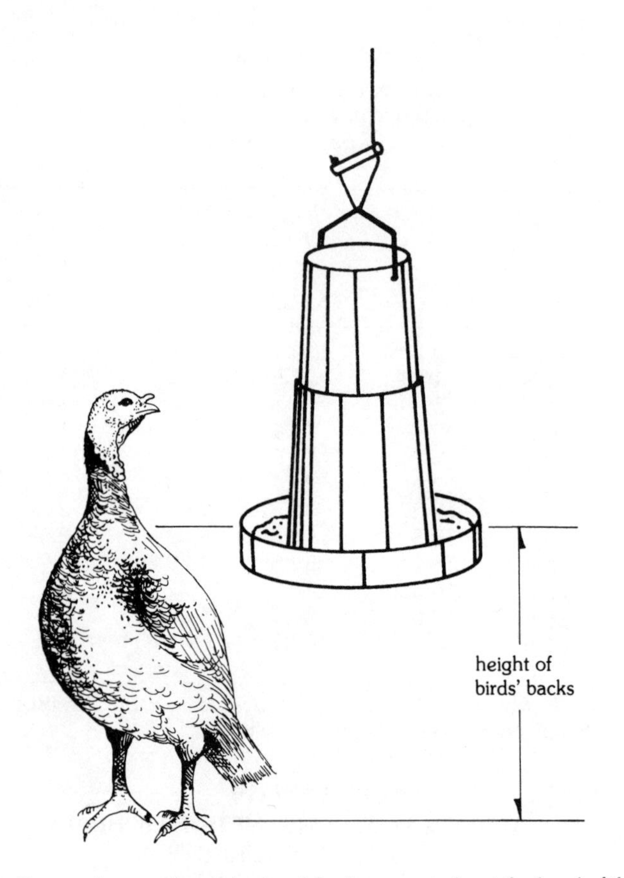

Figure 3-7. To avoid waste, keep the lip of the hopper at about the level of the birds' backs.

the hopper on a level with the birds' backs (Figure 3-7). This means adjusting the feeders if they are of the type that can be adjusted, or changing to larger feeders that will provide this situation. Hanging, tube-type feeders are excellent for brooding or confinement rearing. They may be adjusted easily to prevent waste. One tube feeder 16 inches in diameter is adequate for 25 birds.

CHAPTER 4

The Breeding Flock

Some small flock owners want to hold over some of their turkeys as breeders and produce hatching eggs. The turkey-breeding flock requires time and good management. Not only that, it is an expensive project. The cost of growing breeders to maturity is several dollars (Table 1, page 2) and a considerable amount of feed is required to maintain the flock during the holding period (see Table 5, page 50).

Selecting the Breeders

Start the turkeys to be kept as breeders approximately eight months before egg production is desired. Birds that commence laying too early lay small eggs and small eggs tend to be less fertile.

Keep the best birds, those that are healthy and vigorous, for breeders. Prospective breeders should have good, full breasts (those with non-protruding keel bones). Select also for straight legs, backs and keel bones.

Debeak birds to be placed in the breeding pen; remove about one half of the upper beak. Feed and water levels must be deeper to allow for the shorter beaks.

Mating

For light-type birds, 1 tom per 20 hens is adequate. Normally, medium turkeys are mated in a ratio of 1 to 18 hens. Large turkeys require 1 tom per 16 hens. Keep a few spare toms to replace those that die or are poor mating birds.

TABLE 5
FEED CONSUMPTION OF TURKEY BREEDERS
(Pounds per Bird per Day)

Type Turkey	*Hens*	*Toms*
Large	0.60	1.00
Medium	0.45	0.75
Small	0.35	0.65

SOURCE: *Turkey Production,* Agriculture Handbook No. 393, United States Department of Agriculture.

Lighting

Approximately four weeks prior to mating, stimulate the toms with light. Light stimulation is needed to get good sperm production. The hens should receive light approximately three weeks prior to the onset of egg production.

Thirteen to fourteen hours of light (artificial and natural light combined) is adequate. One 50-watt bulb for every 100 square feet of floor space is adequate. You can use a time clock to bracket the natural daylight hours with artificial light.

Egg Production

Light turkey varieties may be expected to lay 85 to 100 eggs per bird. The medium varieties lay 50 to 70 eggs and the heavies may be expected to lay approximately 50 eggs.

Egg production is best the first year. Thereafter, the egg production rate diminishes by 20 percent each succeeding year. Egg size increases the second year, but hatchability tends to decrease.

Mating Habits

Hens begin to mate when egg production commences. If the toms have been prelighted they are ready to fertilize the eggs effectively. During the breeding period, the strutting or courting activity increases among the males. Then hens select the toms of their choice and squat near them. The male mounts the hen and usually copulation occurs. Some of the males may not mount the females while others may mount them but not complete the insemination. In these cases, the hens may lose their interest and not mate for some time and poor fertility is a problem.

Fertilization

After successful insemination, the sperm make their way up the oviduct to the upper part of the oviduct or funnel. This is the storage area for sperm and the site of fertilization as the ova (yolks) pass through. Sperm are stored in the funnel for several days to several weeks. Storage life of the sperm diminishes with the age of the birds.

Artificial Insemination

In some instances, the hens are fertilized solely by artificial insemination. The toms and hens are kept separate; semen collected from the toms is administered to the hens. In other situations, the hens are inseminated artificially on a supplementary basis to improve fertility where natural mating has not produced good results.

Semen may be collected from the toms two or three times per week. To milk or work a tom requires two people. One holds the bird on a padded table or on his lap with the tail end of the bird toward the other operator. The legs are held slightly spread apart to expose the abdomen. The second person then stimulates the tom by stroking the abdomen and pushing the tail upward and toward the bird's head. The male responds and the copulatory organ enlarges and partially protrudes from the vent. The second person grips the rear of the copulatory organ with his thumb and forefinger from above and fully exposes the organ. The semen is then squeezed out with a short sliding downward movement. The toms become trained quite quickly and ejaculate easily when stimulated. The semen is collected in a small glass beaker or a stoppered funnel. A tom produces 2/10 to 1/2 cubic centimeter (cc) of semen per milking.

It is important that the semen be clear and free of fecal matter. It is possible to avoid some contamination by withholding feed from the toms 8 to 12 hours prior to collecting semen.

Unlike some types of semen, turkey semen cannot be held very long. It should be used within 30 minutes after collection. Good results are obtained when hens are inseminated twice at four-day inter-

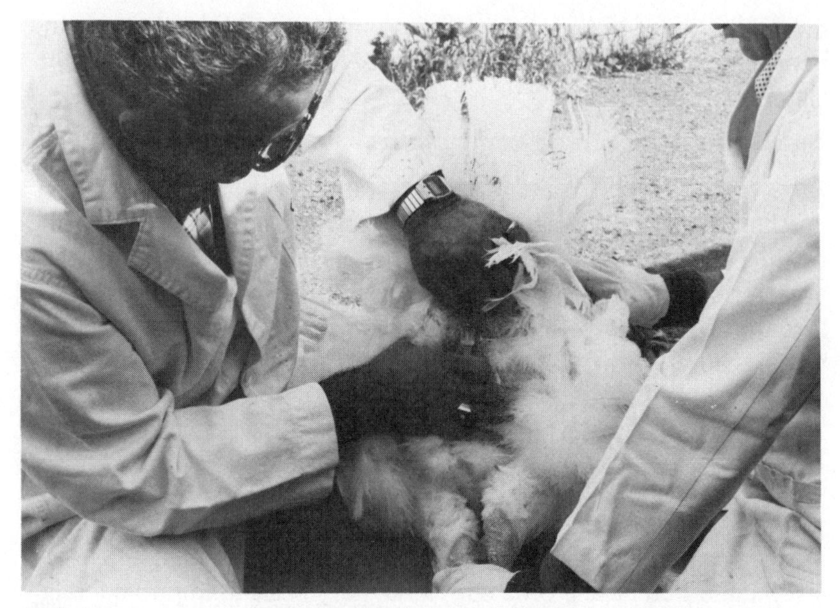

Artificial insemination. One person holds the bird while another stimulates him.

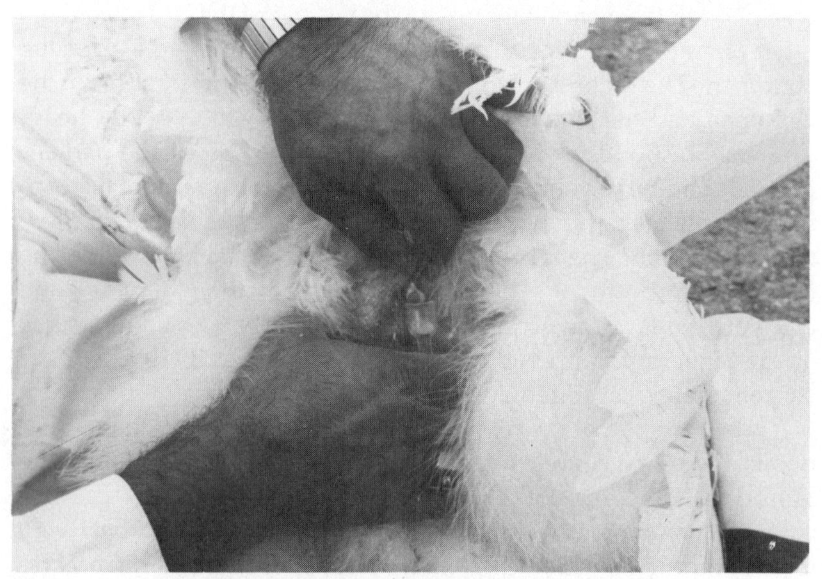

The semen is collected in a small beaker or funnel.

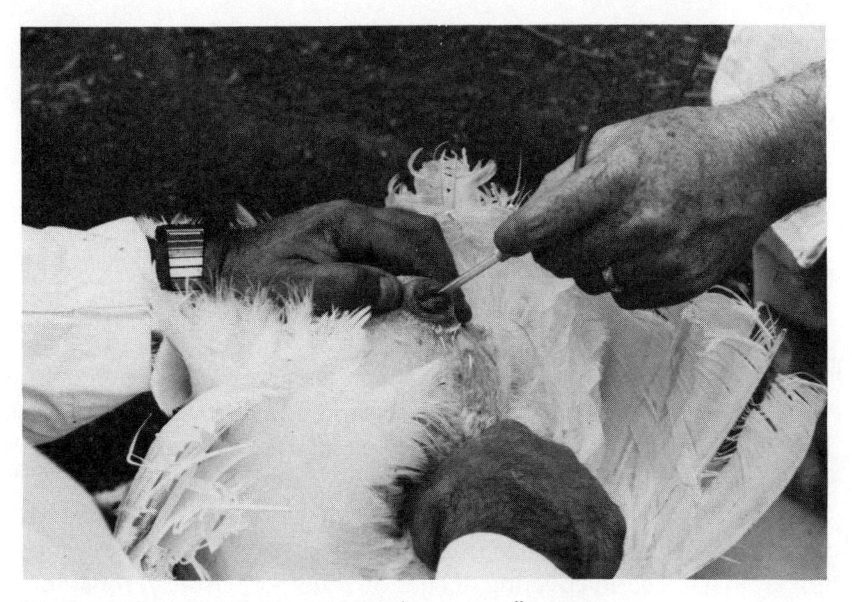

The hen is inseminated using a glass tube or a small syringe.

vals when egg production begins. It should be done every two or three weeks thereafter unless fertility is high in the mated flock.

To inseminate the hens, the opening of the oviduct is exposed and a small syringe without needle inserted into the oviduct about 1½ inches. Special devices such as glass tubes or plastic straws may also be used to inseminate.

Again, inseminating is best done with two people. One sits down with the bird facing him, the breast resting on his lap. He exposes the oviduct by exerting pressure on the abdomen while at the same time forcing the tail upward toward the head. The oviduct can be exposed only in the hens that are in laying condition.

Broodiness

Broodiness is the tendency of the hen to set on, or want to hatch eggs. If broodiness is permitted to go unchecked, egg production will suffer. Turkeys are inclined toward broodiness, an inherited characteristic. Nests should be checked in the early morning or evening for broody birds. Broodys should be removed from the nest immediately and broken of the habit. They should have no nests. Special pens with slat or wire floors are best for broody birds. One or more toms in the pen will tend to keep them on their feet and discourage broodiness.

Broodiness may be discouraged by moving the birds to different areas, by providing roosts and chasing birds off the nest when the eggs are gathered. Collecting eggs several times a day also discourages broodiness.

Housing and Breeders

Buildings used for growing poults can also be used for breeding birds. Although adult turkeys don't necessarily need warm houses, it's best if the house is well insulated and ventilated for maximum comfort during cold or hot weather. Floors can be cement, wood or asphalt. Dirt floors are not recommended because of the potential disease and parasite problems. Clean and disinfect floors thoroughly between flocks. Place a good covering of litter on the floor. Usually, roosts or porches are not used for breeding flocks.

The house needs electrical wiring because stimulatory light for the breeders is needed when days are short.

Breeders require more floor space than growing birds. Usually 6 to 8 square feet per bird is recommended. When hens are housed separate from the toms, 5 to 6 square feet of floor space is adequate for large turkeys and 4 to 5 for small ones.

In areas where the climate is warm or where winters are mild, breeders may be allowed access to a fenced range or yard area. When a range is used, provide 150 square feet of area per bird. A yard should provide 4 to 5 square feet per bird. The shelter should be equipped with roosts and provide at least 4 square feet of space per bird. Locate feeders and waterers and the broody pen in the shelter.

Nests

Make nests available before egg production begins to give the hens an opportunity to become accustomed to them. One nest for every four birds is sufficient. Locate the nests in an area of subdued light.

Open-type nests are satisfactory for the small flock. Trap nests or tie-up nests are available for those who want to do some breeding work. Nests should be 24 inches high by 18 inches wide and 24 inches

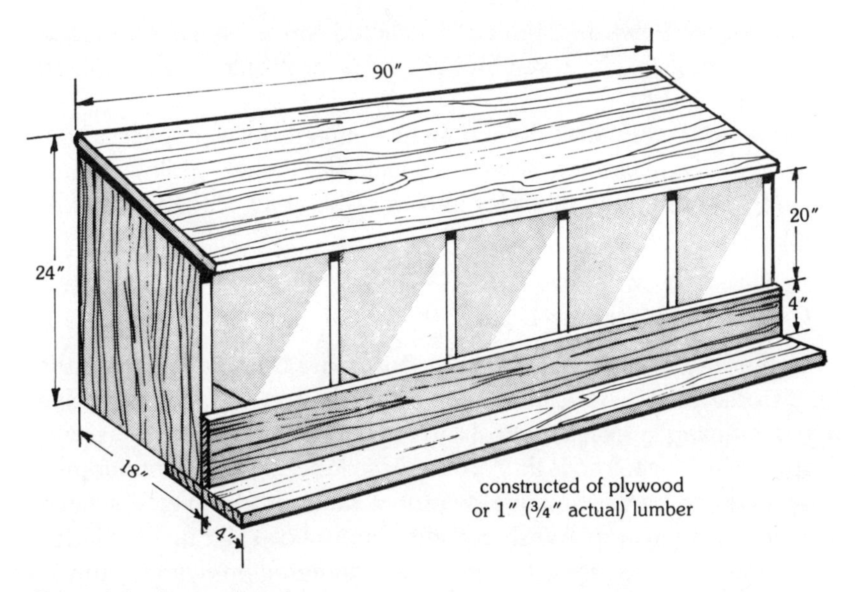

Figure 4-1. A typical poultry nest.

deep. A foot board approximately 5 to 6 inches high on the front will hold the nesting material in the box. Barrels or boxes are sometimes used. If for any reason the nests are outside the shelter, cover them to keep out the weather.

Good nesting materials are shavings, chopped straw, sugar cane and rice hulls. A typical poultry nest is shown in Figure 4-1.

Feeding the Breeders

One month before the onset of egg production, put the breeders on a turkey-breeder diet. Feed according to the instructions of the feed manufacturer. The turkey breeder diet is usually formulated to provide 17 percent protein. All-mash diets are preferred, but mash and grain diets may be used if fed in the right proportions. If too much grain is fed in relation to the mash, egg production and hatchability may be reduced. The mash and grain system has the advantage of making use of home-grown grains without the need for grinding or mixing. Home-grown grains can also be used in an all-mash program by grinding and mixing with a concentrate at home or having them custom mixed elsewhere.

The breeder diet may be fed in pelleted form. There is less feed waste with pellets. An insoluble grit or gravel should be available to confined breeders.

Large turkey breeders need 6 inches of feed hopper space or the equivalent. Smaller birds can get by with 4½ inches. Follow manufacturer's recommendations for tube or other types of feeding equipment.

Care of the Hatching Eggs

Gather the eggs *three times daily* and more frequently if the birds tend to use certain nests more than others. Frequent gathering helps avoid breakage, excessively dirty eggs, possibly frozen eggs and broodiness. Discard eggs that are badly soiled. Slightly to moderately dirty eggs can be washed in a detergent sanitizer formulated specifically for washing eggs. Wash at a temperature of 110° to 115°F. for three minutes. Excessively dirty eggs gathered in warm, humid weather may easily become contaminated, and contaminated eggs frequently explode in the incubator.

A basket egg washer.

The washing job can be done with a basket washer that agitates the water or rotates the egg basket during the washing process. Don't wash the eggs longer than three minutes. After washing, rinse the eggs in water the same temperature as the wash water or higher. The water should contain an approved sanitizer such as one of the quaternary ammonium compounds at a concentration of 200 parts per million (ppm).

If the eggs are held for longer than seven days before incubating, turn them daily. This is done by placing the eggs in a flat or carton and propping one end of the carton at an angle of approximately 30°. Each day, shift the carton so the end to be propped changes every 24 hours. Hatchability improves if this is done. Keep eggs held for more than a day in a storage area at a temperature of 50°-60°F. and a relative humidity of 75 percent. Store the eggs with the large ends up.

Hatching with Broody Birds

Eggs can be hatched by setting them under broody birds. Chickens, turkeys and even ducks or geese can be used. It makes more sense to keep the hen turkeys laying and hatch the eggs by other means. A medium-size broody chicken will cover six to seven turkey eggs. Select a calm bird, one that is not likely to become easily frightened and break the eggs.

Provide a suitable nest box (Figure 4-2), one that is roomy and

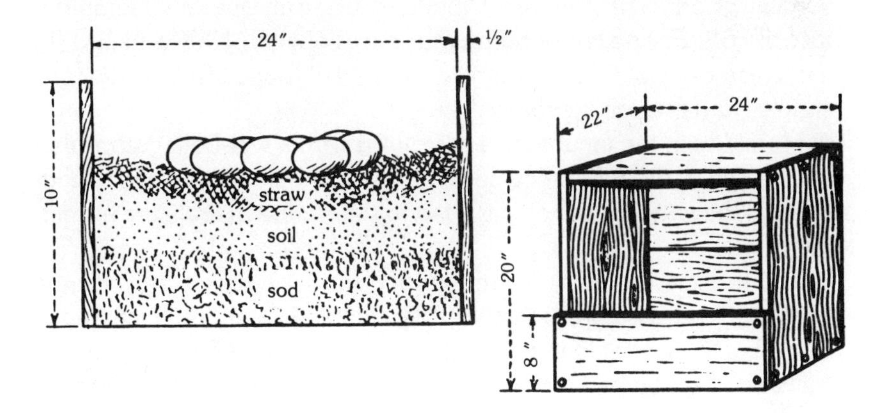

Figure 4-2. The construction of a nest (left), and a nest box (right) for a setting hen.

deep so the setting hen will have ample room to turn the eggs, change position and be comfortable over a 28-day incubation period. Select an area where she will be by herself or other setters. She should have food and water handy at all times.

It may be advisable to put the setting hen on dummy or artificial eggs for a few days to make sure she's a persistent brooder. Try taking her off the eggs a few times. If she immediately goes back on the eggs, put the real things under her. Check broody birds for lice and mites before setting and treat them for these pests if necessary.

Hatching with Artificial Incubators

Two types of incubators are used today. These are the forced-draft machine and the still-air machine. The large commercial hatcheries use forced-draft machines. This type of incubator has fans that force air through the machine and around the eggs. For most types of eggs, the temperature setting of the forced-draft machine is 99.5° to 99.75°F.

The majority of the still-air machines are quite small. The egg capacity ranges from 1 to about 100 eggs. Still-air machines do not have fans but depend upon gravity ventilation through vents on the top and bottom of the machine. The operating temperature of the still-air machine is higher than the forced-draft. It ranges from 101.5° to 102.75°F. depending upon the type of egg being set. Actually, reasonably good results can be obtained by using an operating temperature of 102°F. The temperature may vary between 100° and 103°F. if it doesn't stay at these extremes. Figure 4-3 (page 60) is a plan for a simple home-made incubator.

Humidity in the incubator is measured with a wet-bulb thermometer. Actually, the difference between the dry-bulb temperature reading and the wet-bulb reading is used to determine the relative humidity.

The relative humidity recommended for large turkey eggs during the first 25 days is approximately 62 percent. At a dry bulb reading of 99.5°F., the wet bulb reading should be 87.5°F. After the eggs are transferred to the hatching compartment at around 25 days, the relative humidity should be about 70 percent. With the hatcher running at 99.5°F., the wet bulb reading should be about 90°F.

A forced-draft incubator.

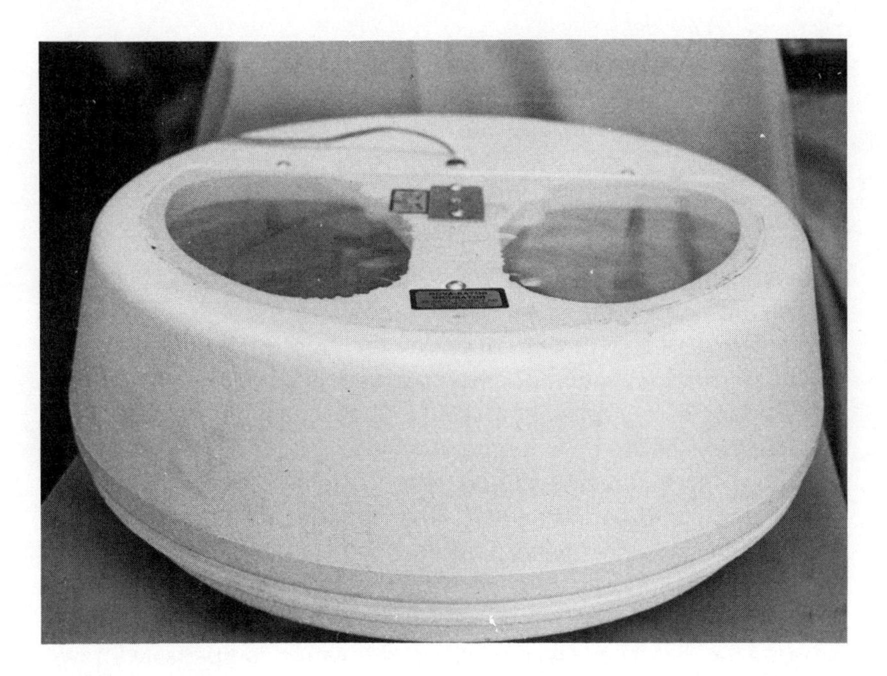

A still-air incubator.

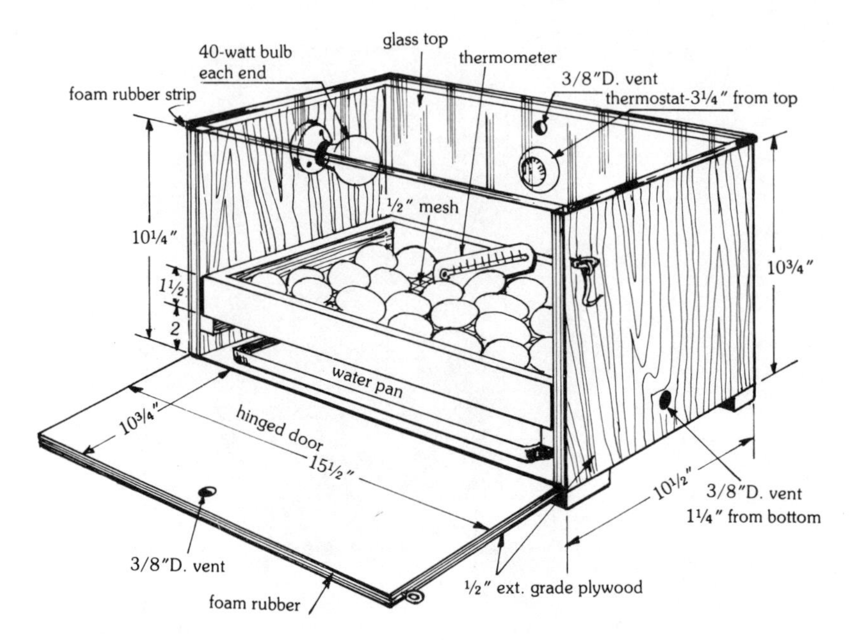

Figure 4-3. A plan for a simple home-made incubator.

Humidity in small machines is usually furnished by evaporation pans. Humidity conditions have to be controlled by the amount of evaporation pan area used or by controlling the ventilation. When adding water to the evaporation pans, use warm water to avoid reducing the incubator temperature. If humidity is too high, open ventilators slightly; if too low, cut down on the amount of ventilation.

The normal incubation period for turkeys is 28 days. Actually, the first 24 days is frequently referred to as the incubation period and the last four days as the hatching period.

During the incubation period (first 24 days), the eggs will lose weight through evaporation at a rather definite rate depending upon the humidity in the incubator. Eggs may be bulk weighed at the time of setting and periodically thereafter as a check on humidity conditions. Table 6 shows the anticipated weight loss at various periods. Humidity in the incubator can be adjusted up or down to compensate for two high or too low a weight loss.

Figure 4-4 shows the approximate size of the air cells when evapo-

TABLE 6
WEIGHT LOSSES DURING INCUBATION

Days of Incubation	*Weight Loss (percent)*
6	2.5
12	5.0
18	7.5
24	10.0

SOURCE: *Turkey Production*, Agriculture Handbook 393, United States Department of Agriculture.

ration is normal. By candling incubating eggs at various stages, this can be used as a rough guide to control humidity in the absence of a wet bulb thermometer.

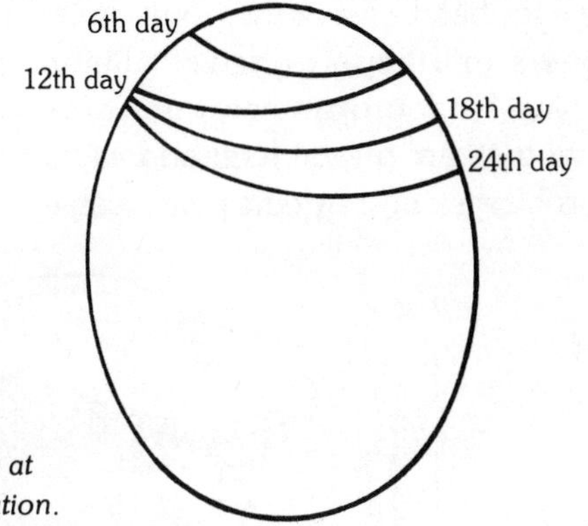

Figure 4-4. Air cell size at various stages of incubation.

Turning the Eggs

During incubation, the eggs should be turned at least four times daily. The purpose of turning is to prevent the embryo from sticking to the shell membrane. Some machines have automatic turning devices on them.

If eggs are packed in the trays large end up, the trays should be slanted about 30°. If the eggs lie flat on their sides, they should be turned 180°. One method of making sure all eggs are turned is to mark them with an X on one side of the egg and an O on the other, in pencil. When turning, keep all the Xs or Os on top. Discontinue turning eggs the last three or four days of the incubation period.

Candling

Frequently with small machines, the eggs are candled three or four days prior to hatching and the infertiles removed. Candling is done in a dark room using a special light. If the eggs are infertile, they will appear clear before the candling light. Fertile eggs incubated for 24 to 25 days permit light through only the air cell or large end of the egg. The remaining part of the egg will appear black or very dark in color. Eggs may be candled at 72 hours to determine fertility. At that point the fertiles will have a typical blood vessel formation, looking much like a spider. Figure 4-5 shows a candling light that can be built at home.

Hatching

When the eggs hatch, leave the poults in the incubator for approximately 12 hours until they are dried and fluffy before removing them. They can survive for approximately three days without food or water, but the sooner they are put on feed and water the better. For a list of incubating problems and remedies, see Table 7.

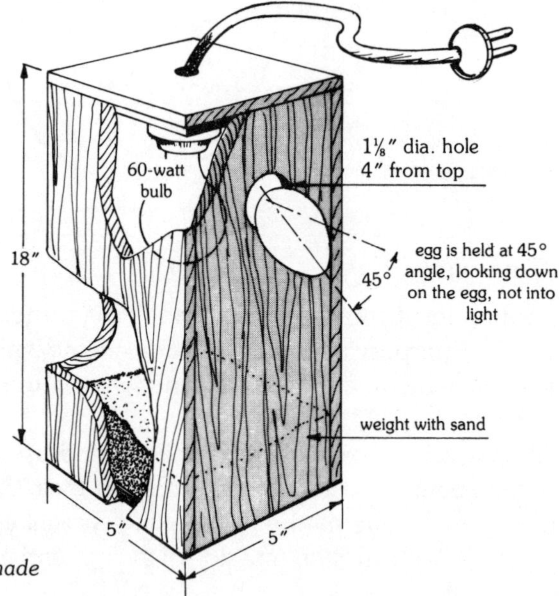

Figure 4-5. A home-made candling light.

TABLE 7
INCUBATION TROUBLE-SHOOTING CHART

Symptom of Trouble	*Probable Cause*	*Suggested Remedies*
1. Eggs clear: no blood ring or embryo growth	1. Males sterile	Careful culling; select for high hatchability.
	2. Males too old	Do not use.
	3. Eggs too old, chilled or overheated	Set within 10 days, collect frequently, store at right temperatures.
	4. Birds too closely confined	Provide adequate floor space.
	5. Seasonal decline in fertility	Use early hatched males, timed for best maturity.
	6. Inadequate nutrition or inadequate water	Feed a breeding diet; provide adequate waterers, well distributed.
2. Eggs appear clear when candled, but show blood or very small embryo when broken out	1. Incubator temperature too high or too low	Check thermometer; operate at correct temperature.
	2. Badly chilled or overheated or held too long	Collect eggs frequently; store at 50° to 55°F., 75% relative humidity.
	3. Breeding flocks out of condition	Do not set eggs from sick, or birds in poor condition or from birds recently vaccinated.
	4. Improper nutrition	Feed breeder ration of good quality.
	5. Poor hatchability heredity	Select strain known to have high hatchability.

TABLE 7—Continued

INCUBATION TROUBLE-SHOOTING CHART

Symptom of Trouble	*Probable Cause*	*Suggested Remedies*
3. Many dead germs	1. Faulty incubator temperature	Check accuracy of thermometer; operate at correct temperature.
	2. Lack of ventilation	Provide plenty of fresh air in incubator room and good ventilation of the incubator.
	3. Improper turning	Turn eggs four times daily.
	4. Low vitamin rations	Feed high-quality breeder ration.
4. Poults fully formed, but dead without pipping; may have considerable quantities of unabsorbed yolk	1. Low average humidity in incubator; too low or too high a humidity at transfer time in hatcher	Maintain proper humidity throughout incubation and hatching cycle.
	2. Temperature, ventilation or improper turning	Follow recommendations on temperature and ventilate room and machine. Turn four times daily.
	3. Chilled eggs	Gather eggs frequently, hold under proper conditions.
	4. Disease or flock in poor condition	Diagnose disease and correct flock problems.
5. Eggs pipped but poults dead in shell	1. Low average humidity	Wet bulb temperature from 85° to 90°F.

	2. High temperature for short time	Maintain recommended temperatures throughout hatch.
	3. Poor ventilation	Provide adequate ventilation of the incubator room and proper openings of the incubator and hatcher ventilators.
	4. Low average temperature	Maintain the recommended temperatures throughout hatch.
6. Sticky poults; poults smeared with egg contents	1. Low average temperature	Use proper temperature.
	2. Average humidity too high	Maintain proper humidity levels.
	3. Inadequate ventilation	Ensure adequate incubator room ventilation and proper adjustment of incubator ventilators.
7. Dry sticks—shell sticking to poults	1. Eggs dried down too much	Ensure proper ventilation and humidity.
	2. Low humidity at hatching	Ensure proper humidity wet bulb 85°F. till pipping—then increase to 88° to 90°F.
8. Poults hatching too early with bloody navels	1. Temperature too high	Maintain proper temperature levels throughout incubation to hatching.
9. Rough navels	1. High temperatures or wide temperature variations	Maintain proper incubator temperature throughout incubation and hatching.
	2. Excessive humidity	Use less humidity first 24 to 36 hours after transfer.

TABLE 7—Continued
INCUBATION TROUBLE-SHOOTING CHART

Symptom of Trouble	*Probable Cause*	*Suggested Remedies*
10. Poults too small	1. Low humidity	Maintain proper humidity levels.
	2. High temperature	Maintain proper temperatures.
	3. Small eggs	Don't set those that are too small.
11. Large soft-bodied poults	1. Low average temperature	Maintain proper temperature.
	2. Poor ventilation	Adequate ventilation of incubator room and incubators.
12. Weak poults	1. Excessive temperature in the hatcher	Watch temperature in hatcher especially after hatch is complete.
	2. Inadequate ventilation in hatcher	Check ventilators; open wider as hatch progresses.
	3. Breeder flock condition	Diagnose and correct flock problem.
13. Short down on poults	1. High temperature	Maintain correct temperature throughout.
	2. Low humidity	Maintain proper humidity.
14. Hatching too early with bloody navels	1. Temperature too high	Maintain proper temperatures throughout incubation and hatch.

15. Draggy hatch; some poults early, but hatch slow in finishing	1. Improper handling of hatching eggs	Gather frequently; store properly.
	2. Improper temperature	Maintain proper temperature throughout.
16. Crippled and malformed poults	1. Cross beak-heredity	Careful flock culling.
	2. Missing eye—rare but may be due to high temperatures	Matter of chance.
	3. Wry neck—nutrition suspected	
	4. Crooked toes	Maintain proper temperature.
	5. Spraddle legs—smooth surfaces in hatching trays	Use crinoline cloth in hatch trays.
17. Malformed poults; Excessive malpositions	1. Eggs chilled	Gather frequently; hold under proper temperature and humidity conditions.
	2. Improper turning	Turn four times daily.
	3. Inadequate ventilation	Provide adequate room ventilation: set ventilator controls on machine.
	4. High or low incubator temperatures	Maintain correct temperatures.
	5. Low humidity	Maintain proper humidity level.
	6. Improper nutrition	Use high-quality breeder diets.

SOURCE: Based on material provided by the Robbins Incubator Co., Denver, Colorado.

CHAPTER 5

Flock Health

Normally, losses from disease are not much of a problem with small turkey flocks. However, there are several diseases that may possibly affect your flock. The old adage that an ounce of prevention is worth a pound of cure certainly applies to growing turkeys. If you purchase stock from a good, clean source, follow a sound sanitation program, use a good feeding program and provide a comfortable growing environment, you will have gone a long way toward keeping your flock healthy.

However, losses will occur on occasion. Commercial flock owners, for example, expect a mortality of somewhere between 3 and 4 percent. So, if you lose one bird and the rest of the flock is eating and drinking and they look healthy, don't get too excited about it. However, a disease in the flock is usually accompanied by a drop in feed or water consumption as well as the appearance of sick or dead birds. When it is apparent that a disease is present, seek the advice of a person who is a trained poultry diagnostician. It is not advisable to use drugs or antibiotics indiscriminately. Sometimes this will do more harm than good and the only result may be a waste of money.

Where there are no local diagnosticians, sample birds may be submitted to a state diagnostic laboratory. The sample should include two or more sick, or recently dead birds. Preserve dead specimens by keeping them cool or freezing them so as to prevent decomposition. An early diagnosis and fast treatment is always recommended as the quickest solution to poultry disease problems. The addresses of the state diagnostic laboratories may be found in the back of the book.

There are several diseases and parasites that may affect turkeys but only the more common ones will be described, and these will not be discussed in great detail. There are many excellent texts on poultry diseases if more in-depth information is needed (see Bibliography).

TURKEY DISEASES

Aspergillosis or Brooder Pneumonia

Aspergillosis is usually a disease of young birds. However, it can affect older birds too. The symptoms are: the birds act dumpy; breathing may be rapid; there may be some gasping and possibly inflamed eyes.

This disease is caused by a fungus which is inhaled by the birds usually from moldy litter or feed. On postmortem, yellowish-green nodules may be found on the lungs in the trachea, bronchi and viscera. There is no treatment known. However, further spread of the infection may be prevented by culling the sick birds and cleaning and disinfecting the house and equipment thoroughly. Moldy litter or feed should be removed carefully from the building to prevent further spread of the disease.

Blackhead or Histomoniasis

Blackhead is caused by a protozoan parasite. It is a very common disease of turkeys of all ages. It can also affect chickens. Since chickens may act as an intermediate host for the organism causing Blackhead, it is somewhat risky to keep chickens and turkeys on the same farm. It is recommended that they not be kept in the same house, and chickens and turkeys should never be intermingled. The term Blackhead is somewhat misleading because that symptom may or may not be present with the disease.

Cecal worm eggs can harbor the organism causing Blackhead over long periods of time. When picked up by the turkeys they infect the intestines and liver. The chicken commonly hosts the cecal worm, thus, the danger of having chickens with turkeys together can be seen.

Mortality with this disease may reach 50 percent if treatment is not started and the infection checked immediately. Symptoms may be droopiness and dark heads, brownish-colored and foamy droppings.

On autopsy, an inflammation of the intestine and ulcers on the liver may be seen. The incidence of the disease and its severity depends upon the management and sanitation programs used. Sanitation of the brooding facilities, rotation of the range areas, and segregation of young birds from old birds help to prevent outbreaks. Segregation of turkeys from chicken flocks is very helpful in preventing problems with Blackhead.

Blackhead drugs are commonly added to the feed to prevent outbreaks and may be used to treat Blackhead infections also.

Fowl Cholera

Fowl Cholera is caused by a bacterium. It is a highly infectious disease of all domestic birds, including turkeys.

The birds become sick rapidly and may die suddenly without showing external symptoms. They may appear listless, feverish, drink excessive amounts of water and show diarrhea.

Post mortem findings include red spots or hemorrhages on the surface of the heart, lungs, intestines or in the fatty tissues. The birds may have swollen livers (a cooked appearance with white spots). Treatment with sulfonamides, such as sulfaquinoxaline or sulfamethazine, is currently recommended. Sulfaquinoxaline in the feed at 0.33 percent level for 14 days is considered to be one of the best treatments. Antibiotics are sometimes injected at high levels. Sanitation in the poultry house, range rotation and proper disposal of dead birds help to prevent Cholera. In problem areas, vaccines can be used and are recommended.

Coccidiosis

Coccidiosis is a very common disease of poultry. It is caused by a protozoan parasite coccidia. The birds expose themselves to the disease by picking up sporulated oocysts in fecal material and litter. It should be assumed that all flocks grown on litter are exposed to the disease. Birds grown on elevated wire or slats are not exposed to droppings and normally don't become exposed to Coccidiosis.

Coccidia are host specific, that is, the coccidia that affect turkeys do not affect chickens. Different species affect various parts of the digestive tract. Six species are known to infect turkeys, but only two of these are commonly troublesome.

Symptoms of Coccidiosis are ruffled feathers, head drawn back into the shoulders and the appearance of being chilled. In the industry, birds having this appearance are sometimes called *unthrifty*. A bloody diarrhea may be seen with some forms. If permitted to go unchecked, the turkeys may die.

On post mortem, lesions and hemorrhages may be seen in the various parts of the intestine, depending upon the species involved. The disease may be prevented by feeding coccidiostats in the starter diet at low levels and permitting the bird to build an immunity, or completely controlling the disease by feeding a preventive level of coccidiostats in the diet.

Treatment involves the use of sulfonamides or other coccidiostats as prescribed by a diagnostician or service person.

Erysipelas

Erysipelas is caused by a bacterium. Swine, sheep and man, among others, are also susceptible to the disease.

Symptoms in turkeys include swollen snoods, bluish-purple areas on the skin and congestion of the liver and spleen. Birds may become listless, have swollen joints and exhibit a yellow-green diarrhea. It is primarily a disease of toms because the organism readily enters through wounds caused by fighting. Since the snood is frequently injured when toms fight, this is a common site for erysipelas infection. For this reason, some commercial producers have their turkeys desnooded at the hatchery or remove snoods on the farm at day-of-age. Erysipelas is a soil borne disease and contaminated premises are the primary source of infection. Control requires good management and sanitation. The disease responds well to penicillin. Tetracycline is also effective. Vaccination is recommended in those areas where the disease is common. If this disease is suspected, use care. Wear gloves when performing an autopsy on a diseased bird.

Fowl Pox

Fowl Pox is found in many areas of the country. It is caused by a bacterium and spread by contact with infected birds, or by such vectors as flies, mosquitoes or wild birds. There are two forms of fowl pox—the dry or skin type and the wet or throat type.

Birds with Fowl Pox have a poor appetite and look sick. The wet pox causes difficult breathing and nasal or eye discharge, yellowish-soft cankers of the mouth and tongue. The dry pox causes small grayish-white lumps on the face. These eventually turn dark brown and become scabs.

On post mortem, cankers may be found in the membranes of the mouth, throat and windpipe. There may be occasional lung involvement or cloudy air sacs. There is no treatment for the disease itself, although an antibiotic may help to reduce the stress of the disease.

The only means of control is by vaccination. This is recommended in areas where Fowl Pox is a problem.

Newcastle Disease

Newcastle disease is widespread. It is acute, highly contagious and found in chickens, turkeys and other species of poultry. It is a respiratory disease caused by a virus. It causes high mortality in young flocks. In breeder flocks, egg production frequently drops to zero.

Newcastle spreads rapidly through the flock causing gasping, coughing, and hoarse chirping. Water consumption increases and a loss of appetite occurs. Infected birds tend to huddle and exhibit signs of partial or complete paralysis of the legs and wings. The head may be held between the legs or on the back with the neck twisted.

The disease is transmitted in many ways: it can be tracked in by people, brought in with other birds from another site or on dirty equipment, feed bags or by wild birds that gain entrance to the pen.

Post mortem findings may include congestion and hemorrhages in the gizzard, intestine and proventriculus. The air sacs may be cloudy.

There is no effective treatment, though antibiotics are normally given to hold down secondary invaders. Vaccination is recommended in most areas of the country and can be administered individually or on a mass basis. Newcastle disease vaccines can be administered intranasally, ocularly or by wing web. (The wing web is the thin layer of skin at the foreward edge of the wing between the forearm and the tip of the wing or hand.) On a mass basis, it can be given to the birds in drinking water or in the form of a dust or spray. Follow manufacturer's recommendations for use of these products. The vaccination program should be that which is recommended for the given area.

Omphalitis

Omphalitis is caused by a bacterial infection of the navel. It occurs when the navel doesn't close properly following hatching. It may be caused by poor incubator or hatchery sanitation, chilling or overheating.

Birds with Omphalitis are weak and unthrifty and tend to huddle together. The abdomen may be enlarged and feel soft and mushy. The navel is infected. The area around the navel may be a bluish-black. Mortality may be high for the first four to five days.

There is no treatment for the disease. Most of the affected poults die the first few days. No medication is needed for the survivors.

Paratyphoid

Paratyphoid is an infectious disease of turkeys and some other birds and animals. It is caused by one or more of the Salmonella bacteria. Transmission may be from the hen through the egg and to the chick. The organism is also found in fecal material of infected birds.

The disease is primarily one of young birds but older birds may also be affected. In young birds, mortality can run as high as 100 percent. Some birds may die without showing symptoms while others show signs of weakness, loss of appetite, diarrhea and pasted vents. Birds may appear chilled and huddle together for warmth. In older birds, there is a loss of weight, weakness, and diarrhea.

On post mortem of young birds, there may be unabsorbed yolk sacs, small white areas on the liver, inflammation of the intestinal tract, congestion of the lungs and enlarged livers. Older birds usually show no lesions, although a few may show white areas on the liver.

Sulfamerazine, nitrofurans, and some of the antibiotics may reduce losses, prevent secondary invaders and increase the bird's appetite.

Control is through sanitation and isolation of the flock from sources of infection, such as wild birds, birds from other flocks, and contaminated feed and equipment.

Pullorum

Pullorum is an infectious disease of chickens, turkeys and some other species. It is found all over the world. Pullorum is fatal to birds under two weeks of age. It is caused by a bacterium Salmonella pullorum.

Mortality first occurs at five to seven days of age. The birds appear droopy, huddle together, act chilled, and may show diarrhea and pasting of the vent. Pullorum is sometimes called *white diarrhea*.

Salmonella pullorum spreads mainly from the hen to the poult through the egg. It spreads rapidly through transmission on the down of poults located in incubators and hatchers.

Post mortem findings in infected chicks include dead tissue in the heart, liver, lungs, and other organs and an unabsorbed yolk sac. The heart muscle may be enlarged and show grayish-white nodules. The liver may also be enlarged, appear yellowish-green and become coated with exudate. One of several types of blood tests helps to establish a positive diagnosis for Salmonella pullorum.

Sulfamethazine and sulfamerazine are effective in reducing mortality. Nitrofurans may also be effective but will not cure the disease. Medication may hinder diagnosis of Salmonella pullorum.

Flocks that have recovered from Salmonella pullorum should not be kept for replacements or breeding purposes unless they have been blood tested and found to be free of the disease.

Buy poults from Pullorum-Typhoid-free hatcheries.

Infectious Sinusitis or Micoplasma Gallisepticum

Infectious Sinusitis is a disease of turkeys caused by the same organism that causes chronic respiratory disease (CRD) in chickens. It is transmitted through the egg from carrier hens. Stress is thought to lower the poult's resistance to the disease.

Symptoms are nasal discharge, coughing, difficult breathing, foamy secretions in the eyes and swollen sinuses, accompanied by a drop in feed consumption and loss of body weight. Air sac infection may be in evidence on post mortem.

Antibiotics and antibiotic vitamin mixtures will help. Individual treatment with injectable penicillin and streptomycin in the sinuses is recommended.

Infectious Sinovitis or Micoplasma Synoviae

Sinovitis is an infectious disease of turkeys caused by a pleuropneumonia-like organism (PPLO). At first, it was identified as a cause of infections of the joints but more recently it has been known to cause respiratory disease as well. The disease can affect birds of all ages.

Symptoms include lameness, hesitancy to move, swollen joints and foot pads, loss of weight and breast blisters. Some flocks show respiratory symptoms. Dying birds show a greenish diarrhea. Upon post mortem examination, swelling of the joints and a yellow exudate, especially in the hock, wing and foot joints, is in evidence. Internally, the birds may show signs of dehydration and enlarged livers and spleens. Birds showing the respiratory involvement are not easy to spot. On post mortem examination the air sacs may be filled with liquid exudates.

The most common means of transmission is through infected breeders. Poor sanitation and management practices also contribute.

Antibiotics yield some results. These should be given by injection or in the drinking water. Some prefer to use both simultaneously for the best results.

Fowl Typhoid

Fowl typhoid is caused by a bacterium Salmonella gallinarum. It affects chickens, turkeys and other species of birds. The disease may be present wherever poultry is grown.

Affected birds may look ruffled, droopy and unthrifty. Other symptoms may include loss of appetite, increased thirst and a yellowish-green diarrhea. On post mortem the liver may have a mahogany color. The spleen may be enlarged and there may be pinpoint necrosis in the liver and other organs.

Prevention, treatment and control are the same as for Pullorum disease. Buy Pullorum-Typhoid-free poults.

TURKEY PARASITES

There are several parasites that can affect poultry, but relatively few of them are of major importance. Some parasites live outside the bird while others live inside. Certain external parasites can affect growth and performance. Some internal parasites may cause setbacks in weight gain, a loss of egg production in laying birds and even deaths if infestation is severe. Some intestinal parasites harbor other disease organisms that may be harmful to turkeys.

INTERNAL PARASITES

Large Round Worm

Light infestations of round worms are probably not too serious. However, when the worms become numerous, birds can become unthrifty and feed conversion and weight gain suffer. The worms by themselves infrequently cause mortality but when present with other diseases may cause an increase in mortality.

The large round worm is 1½ to 3 inches long. It is found in the upper to middle portion of the small intestine and can cause setbacks in weight gain and a loss of egg production in breeding birds. If the birds have heavy infestations, the worms may be found the full length of the small intestines. Piperazene compounds are used to treat round worms and can be given in the water, the feed or in capsule form.

Ranges that have been used over a period of years, houses with dirt floors or houses that have not been properly cleaned and disinfected are sources of the round-worm eggs. Turkeys infect themselves with the round worms by picking up the egg from the feces or from contaminated ranges or quarters.

Cecal Worms

Cecal worms are very small and by themselves are not injurious. However, they are significant since they act as carriers for the organism causing Blackhead, a very serious disease of turkeys.

Gapeworm

The gapeworm attacks the bronchi and trachea. It can cause pneumonia, gasping for breath, or even suffocation. High mortality may be common among young birds. The treatment of choice is tramisole at 1.6 milligrams per pound of body weight, for three days, given in the drinking water. The earthworm is an intermediate host. The gapeworm is y-shaped and red in color. Small turkeys open their mouths with gaping movement and may have bloody saliva. The worms often can be seen by opening a turkey's mouth and looking down the windpipe.

Tapeworm

There are several species of tapeworms varying in size from microscopic to 6 to 7 inches in length. They are flat, white and segmented. They inhabit the small intestine and cause a loss of weight and lowered egg production.

Effective control of internal parasites depends primarily upon a program of cleanliness and sanitation. Parasitic eggs can remain viable in the soil for more than a year. This means it's important to rotate poultry runs or yards. Preferably, poultry ranges should be used only one year and left idle for three before they are used again. Poultry yards and runs should be located in well-drained areas and be kept as clean as possible. Cultivating and seeding down these areas will help prevent the birds from picking up parasite eggs. The tapeworm needs intermediate hosts like worms, snails or beetles in which to complete part of its life cycle. Turkeys get tapeworms by eating the infected worms, beetles or other hosts.

EXTERNAL PARASITES

Lice

Lice are chewing or biting insects that cause birds considerable grief. With severe infestations, growth and feed efficiency suffer. Egg production in producing birds is also affected. Lice can cause an irritation to the skin with a scab formation.

Lice spend their entire lives on the birds. They will die in a matter of a few hours if they leave the bird. The eggs are laid on the feathers where they are held with a glue-like substance. The eggs hatch in a few days to two weeks. Lice live on the scale of the skin and feathers. Several types attack poultry. Some are gray; others are yellow; it's difficult to distinguish their colors. The body louse, one of the most common poultry lice, usually affects older birds. The lice and its nits (eggs) are seen on the fluff, the breast, under the wings and on the back.

Materials that may be used to treat lice are either *Sevin* (carbaryl), malathion, or *Co-Ral* (coumaphous). Use these materials according to the directions on the label. Examine birds frequently for signs of reinfestation.

Mites

The Northern Fowl Mite. The Northern Fowl Mite is reddish, dark brown. It is found around the vent, tail and breast. These mites live on the birds at all times. They attach to feathers, suck blood, cause anemia, weight loss, and reduced egg production. Materials recommended for treatment of Northern Fowl Mite are *Sevin* or *Co-Ral*. Use according to the manufacturer's directions.

Red Mites. Red Mites are not found on the birds during the day; they feed at night. They may be seen on the underside of roosts, cracks in the wall or seams of the roosts during daylight hours. Other signs are salt and pepper-like trails under roost perches or clumps of manure. Red mites are bloodsuckers. They cause irritation, loss of weight, reduced egg production and anemia. Treat the red mite with the same materials as those used for Northern Fowl Mite.

PRECAUTIONS FOR DRUG AND PESTICIDE USE

Use all drugs and pesticides according to the label directions and use them with caution. To maintain a healthy flock and to obtain optimum production, it is sometimes necessary to administer drugs or pesticides. However, before they are used, make certain of the problem and then use the material of choice. Never use a material that is *not* registered for use on poultry.

There is no substitute for good management. And drugs or pesticides are not intended as substitutes. They work best when combined with good sanitation and sound management practices. Early diagnosis and treatment of a disease or parasite problem is important.

Agencies of the federal government have established definite withdrawal periods and tolerance levels for various materials used in poultry production. For example, the Food and Drug Administration insists that certain drugs be withdrawn a specified number of days prior to poultry slaughter. These withdrawal periods vary from one or two days up to several days and are subject to change. The FDA also establishes maximum limits for residues of certain materials. Some insecticides can be used around poultry but not directly on the birds or on the eggs, or in nests. Some drugs and pesticides cannot be used within a certain number of days prior to slaughter. With some materials, frequency of use may be restricted.

Withdrawal periods and tolerances, as mentioned, *do change* and accepted treatment materials change, so specific precautions for various materials will not be dealt with here. The point is that drugs and insecticides must be used discriminately. Follow all the precautions on the label. Improperly used, drugs and insecticides can be injurious to man, animals and plants.

Store all drugs and pesticides in the original containers in a locked storage area. Keep them out of reach of children and animals. Avoid prolonged inhalation of sprays or dusts and wear protective clothing and equipment when recommended. *BE SAFE*.

NUTRITIONAL DISEASES

A number of poultry diseases may be caused by nutritional deficiencies or imbalances. With today's well-formulated diets, nutritional problems are very infrequent. Therefore, a thorough discussion of nutritional diseases will not be undertaken here. However, to underscore the importance of good nutrition, I'll mention a few of the more common nutritional diseases.

Rickets

Rickets is caused by a deficiency of Vitamin D_3. Birds with rickets appear to be weak, have stiff swollen joints, soft beaks, soft leg bones and enlarged ribs. A deficiency of phosphorus can also cause rickets.

Perosis

Perosis, sometimes called *slipped tendon*, is a leg problem usually caused by a deficiency of manganese in the diet. Heredity may also be a factor. Choline, Niacin and Biotin are thought to also play a role. The large tendon of the leg at the rear of the hock slips to one side resulting in a twisted leg. Once the victim is crippled permanently, it should be killed. With this deficiency, most turkeys will respond to early use of additional manganese in the feed.

MISCELLANEOUS PROBLEMS

Pendulous Crop

When in its normal position, the turkey's crop is in the wishbone cavity and attached to the side and back of the neck. If for some reason the connective tissues that hold the crop in place weaken, the crop drops. If the crop gets too far out of its normal position, feed cannot pass from the crop to the gizzard and the birds actually starve with the crop full of feed. Young birds with a mild pendulous crop condition usually recover. Seriously affected birds seldom recover and treatment is ineffective.

Leg Weaknesses

Leg weaknesses other than Perosis may be caused by vitamin deficiencies or by such diseases as infectious Sinovitis. In day-old poults, a condition resembling perosis is called *spraddle legs*. This may be due to a genetic factor, faulty incubation or a deficiency in the diet of breeding stock. Smooth, slippery surfaces in the hatching trays, shipping boxes or under the brooders may also cause the problem. Place young poults on wire, a rough surface paper (if paper is used) or on litter. Crooked toes may be a hereditary defect or due to faulty brooding conditions.

CHAPTER 6

Killing and Processing

Turkeys should be finished and ready for processing at around 22 to 24 weeks of age. Turkey broilers or fryer-roasters may be processed at 12 to 16 weeks. The age will depend upon the turkey variety and strain, the feeding program and other factors. To determine whether a bird is prime and ready to be processed, check it for freedom of the short pin feathers. The bird is "ready" when the feathers are easy to remove. These are immature feathers that are nonprotruding or may have just pierced the skin. The short protruding feathers have the appearance of a quill with no plume. They are unattractive, particularly in those varieties with dark feathers, and cause downgrading when present in finished market birds. If the bird is not going to be marketed but consumed at home, the pin feathers may stay in place. When the presence of pin feathers is considered to be important it is best to delay dressing those birds until the feathering improves.

Also, check the degree of fat covering. To do this, pull a few feathers from the thinly feathered area of the breast, at a point about half way between the front end of the breast bone and the base of the wing. Take a fold of skin between the thumb and forefinger of each hand and examine for thickness and coloration. On a prime turkey, the skin fold will be white or yellowish white and quite thick. Well-fattened birds will have thick, cream-colored skin while underfattened birds will have thin, often paper-thin, skin that is semitransparent and tends to be reddish.

CARE BEFORE KILLING

Careless handling can cause birds to pile and trample each other, resulting in injuries. Recently injured birds may appear red at the bruise site while old injuries are bluish green. Such defects detract from the dressed appearance. When catching the birds, grab the legs between the feet and hock joints with one hand. Straighten the legs to lock the hock joints. Don't grasp the legs at the feathered area above the hock joints as this may cause skin discoloration. After catching the bird by the shanks, hold one wing with the other hand at the base. This immobilizes the bird effectively. It also gives the handler control of the bird and prevents injuries and bruising.

Starve the birds for approximately 12 hours before killing them. Do not withdraw water to avoid excessive dehydration. Removing the feed enables the crop and intestines to empty before killing. Starving the birds makes the job of eviscerating much cleaner and easier. Remove birds to be starved from the pen and put them into coops containing wire or slat bottoms so they do not gain access to feed, litter, feathers or manure. After catching the birds, keep them in a comfortable, well-ventilated place prior to killing. Overheating or lack of oxygen can cause poor bleeding and result in bluish, discolored carcasses.

PROCESSING AREA AND EQUIPMENT

Home processing of just a few birds requires little in the way of special facilities or equipment. If a fairly large number are to be dressed, it is desirable to have an adequate area and some special equipment such as a mechanical picker.

Process your poultry in as sanitary a manner as possible. It is important to prevent contamination of the carcasses. One of the most common sources of contamination is the contents of the intestine. Contamination which can also come from dirty facilities, equipment or people, results in poorer quality and shorter shelf life, that is, the period before spoilage begins.

Process the turkey in an area that is clean, well lighted, has a water supply, and is free of flies. Helpful equipment includes easily-cleaned working surfaces for eviscerating and packaging plus suitable containers for handling the offal or waste by-products.

At best, the processing job is a messy one. Ideally, there should be two rooms available for processing. Use one room for killing and picking the birds and the other for finishing, eviscerating and packaging when several birds are to be done at one time. If this is not possible or just a small number of birds is involved, then do the killing and plucking in one operation. And then clean the room, draw the birds and package them as a second operation. This two-step method will make the whole procedure far more sanitary.

REQUIRED EQUIPMENT

Shackles or Killing Cones

If only a few birds are to be dressed, a shackle for hanging can be made from a strong cord with a block of wood, 2 × 2 inches square, attached to the lower end. A half hitch is made around both legs and the bird is suspended upside down. The block prevents the cord from pulling through.

Commercial and semicommercial dressing plants use wire shackles that hold the legs apart and make for easy plucking. Some producers make their own shackles out of heavy-gauge wire. Other people pre-

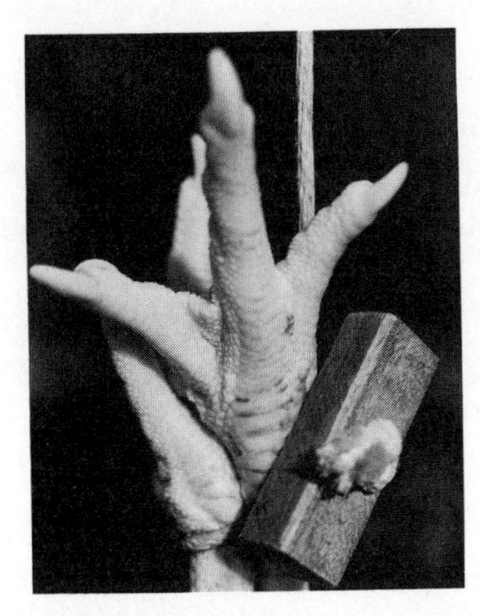

Wood and rope shackle.

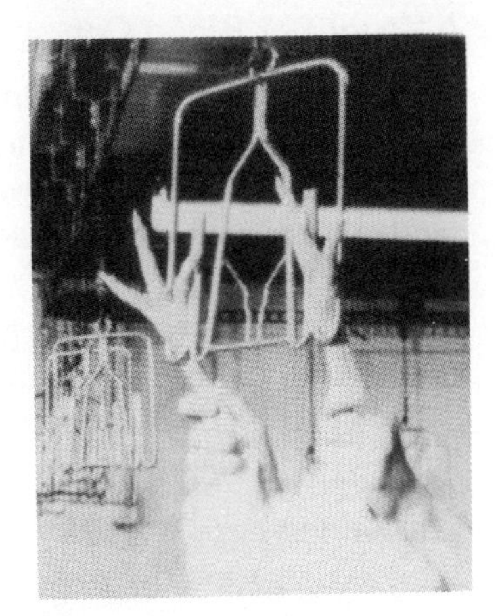

Wire shackle.

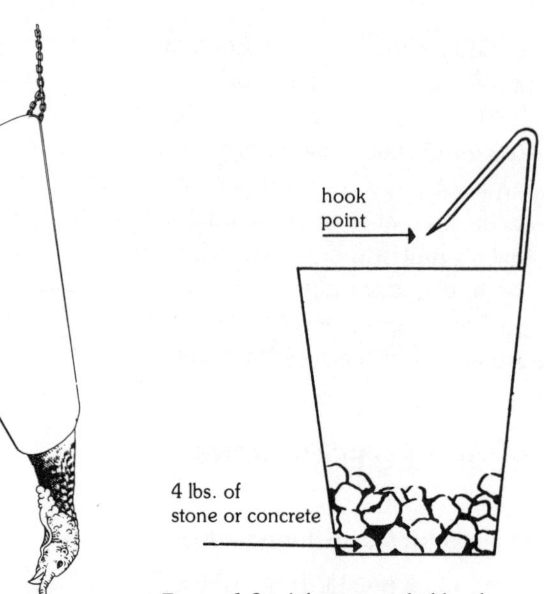

Figure 6-1. Killing cone.

Figure 6-2. A home-made blood cup (not needed with a killing cone).

fer to use killing cones (Figure 6-1) that are similar to funnels. The bird is put down into the cone with its head protruding through the lower end. This restrains the bird and prevents some of the struggling that may lead to bruising or broken bones.

A Weight or Blood Cup

A weighted blood cup or a simple weight attached to the lower beak of the bird will prevent it from struggling and splashing blood. You can make the weight from a window weight attached to the lower beak by means of a sharp hook. The blood cup is not used when killing funnels are available (Figure 6-2).

You can make a blood cup from a two-quart can. Solder a sharp-pointed heavy wire to the can. The wire hooks through the lower beak. Weight the cup with concrete or heavy stones.

The Knives

About any type of knife is satisfactory for dressing poultry. There are special knives for killing, boning and pinning. Six-inch boning knives work very well. If the birds are to be brained, then use a thin sticking or killing knife.

Scalding Tank

When the birds are to be scalded and only a few birds are to be dressed, a 10- to 20-gallon garbage can, or any other clean container of suitable size, is satisfactory. When considerable dressing is done, a thermostatically-controlled scalding vat is preferred. In the absence of the automatically-controlled scalding vat, hot water can be continually heated and the vat replenished as required to maintain a desired temperature. A home-made, electrically heated scalding vat with thermostatic control is shown in the photo below.

Thermometer

Accurate temperatures are important for certain types of scalding. Acquire a good, rugged dairy thermometer, a candy thermometer or a floating thermometer that accurately registers temperatures between 120° and 150°F.

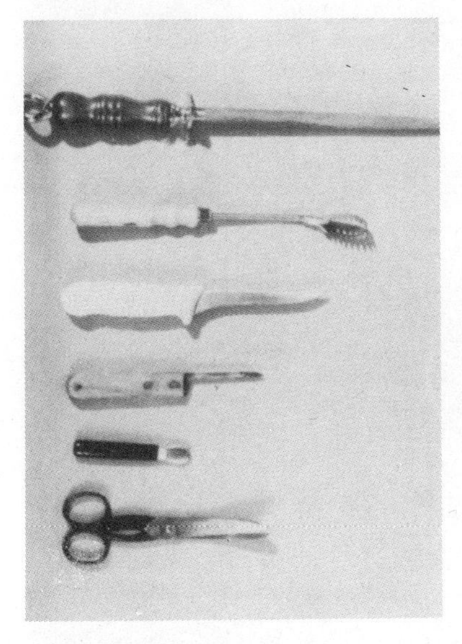

Common poultry knives and implements.

A simple thermostatically controlled scalding tank.

KILLING

Suspend the turkey by its feet with the rope or metal shackle or place it in a killing cone. Hold the head with one hand and pull it down for slight tension to steady the bird. With a sharp knife, sever the jugular vein just back of the mandibles. This can be done by inserting the knife into the neck close to the neckbone, turning the knife outward, and severing the jugular. It may also be done by cutting from the outside.

Another slightly more difficult method is to cut the jugular vein from inside the mouth. Hold the head in one hand, the fingers grasping the sides; take care not to squeeze the jugular veins on the side of the neck. Make a strong deep cut across the throat from the outside close to the head so that both branches of the jugular vein are severed cleanly at or close to the junction. *Warning:* hold the head so your fingers do not get in the way.

Do not grasp the wings or legs tightly so as to restrict the flow of blood from these parts. A poor-appearing carcass will result if bleeding is not complete.

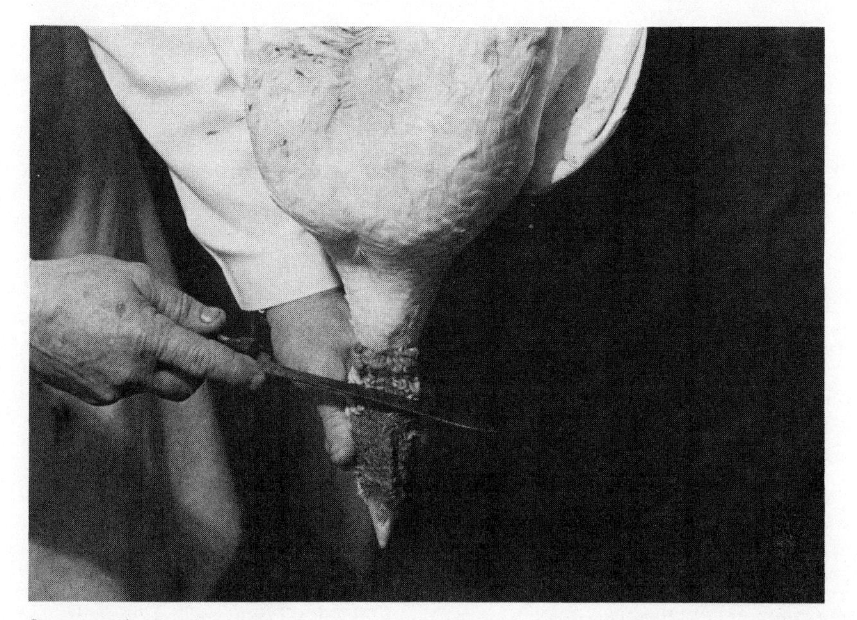

Severing the jugular vein.

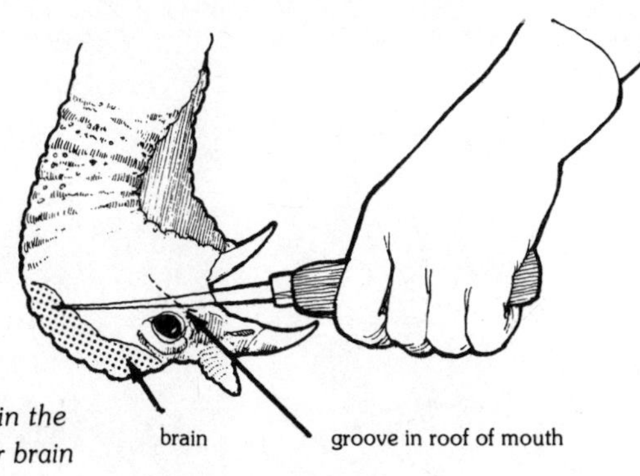

Figure 6-3. To debrain the turkey, pierce the rear brain lobe as shown.

Debraining

Debraining loosens the feathers so that it is easier to pluck the birds. It is done after the jugular vein is cut. Debraining is done when the birds are to be dry picked, but may also be done when the birds are to be semi or subscalded to make the removal of feathers even easier.

To debrain the bird, insert the knife through the groove or cleft in the roof of the mouth and push it through to the rear of the skull so it pierces the rear lobe of the brain (Figure 6-3). Then give the knife a one-quarter turn. This kills the bird and loosens the feathers. A characteristic squawk and shudder indicates a good stick. This procedure requires considerable practice before proficiency is achieved. Though dry picking is slower, the outer layer of skin is not removed making for a fine-appearing dressed carcass.

SCALDING AND PICKING

As soon as the bird is dead and bleeding is complete (usually two to three minutes), loosen the feathers using the subscald method. Dunk the bird in water at approximately 140°F. for about 30 seconds. When one is available, the feathers may be picked by a rubber-fingered picking machine that also removes the cuticle or bloom. The bloom is the thin, outer layer of the skin.

Pin feathers that are left are removed by hand. Don't permit the skin to dry out or it will become discolored. If not immediately eviscer-

Scald the bird in 140°F. water for 30 seconds. Dunk it well.

ated, put the birds in cold running water. The subscald makes it easy to remove the feathers and gives a uniform skin color. However, the skin surface tends to be moist and sticky and will discolor if not kept wet and covered. For the scald to be effective, slosh the bird up and down in the water to get the water around the feather follicles at the base of the feathers.

Semiscald

Another method sometimes used is the semiscald. The bird is scalded for about 30 to 60 seconds in water 125° to 130°F. With the semiscald, the feathers loosen, but the temperature is not hot enough to destroy the outside layer or skin cuticle. Thus the carcasses appear more like dry, picked birds. The water temperature must be within the narrow 125°-to-130°F. range. Time is also a factor and depends upon the age of the bird. If the water is a little cool or the scalding time too short, the feathers will not be loosened enough to pick easily. If feathers pull too hard, skin tears can result. If the water is too hot or the scalding time too long the bird will have an overscalded or patchy appearance.

Hand Picking

If the bird is to be hand picked, rehang it on the shackle and, with a twisting motion, remove the large wing and tail feathers first. Then remove the remainder of the feathers as quickly as possible in small bunches so as not to tear the skin.

Pinning and Singeing

Pin feathers, the tiny immature feathers, are best removed under a slow stream of cold tap water. Use a slight pressure and a rubbing motion. Those that are difficult, you can remove with a pinning knife or a dull knife. By applying pressure between the knife and the thumb, you can squeeze the pin feathers out. The most difficult may have to be pulled. Usually, turkeys have a few hairlike feathers left following hand plucking. You can singe these hairs with an open flame. A small gas torch works well (see below). Take care not to apply the flame directly to the carcass to avoid scorching the skin.

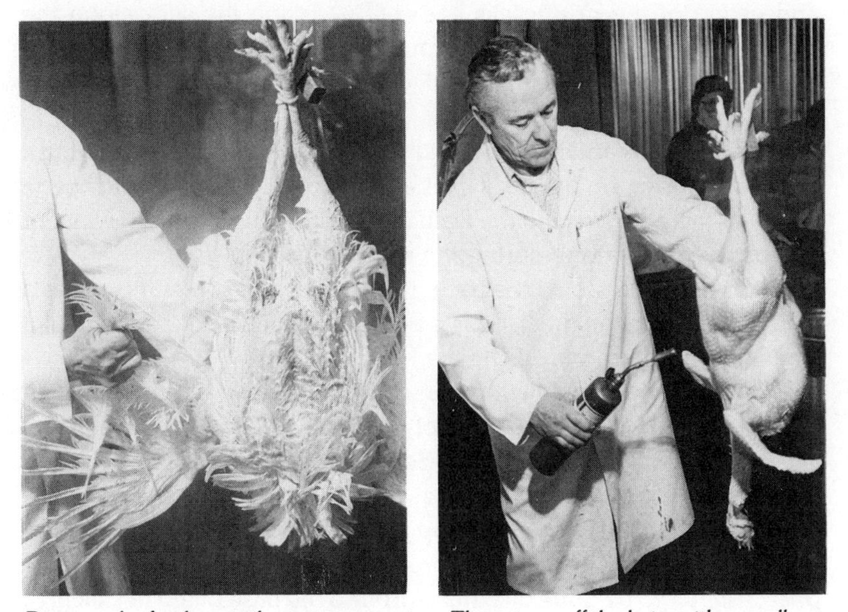

Remove the feathers with a twisting motion. Then singe off the hairs with a small gas torch.

EVISCERATING

After picking and singeing, wash the carcasses in clean, cool water. As soon as they are washed, they are ready for evisceration. Some prefer to cool the poultry first because, after cooling, eviscerating is somewhat easier and cleaner. Others eviscerate and then place the birds in ice water or cool water that is constantly replenished. There are many methods of drawing poultry. The following methods are not the only ones, but are satisfactory. Neatness and cleanliness are essential.

Tools needed for drawing poultry are a sharp, stiff-bladed boning knife, a hook (if the leg tendons are to be pulled) and a solid block or bench upon which to work. Keep all equipment and working surfaces clean. A piece of heavy parchment paper or meat paper may be laid on the working surface and changed as necessary.

Sometimes the tendons are removed from the drumsticks before removing the shanks and feet. Removal of the tendons makes carving and eating of the drumsticks easier. By cutting the skin along the shank, the tendons extending through the back of the leg may be exposed and twisted out with a hook or a special tendon puller if one is available.

Cut the shanks and feet off straight through the hock joint leaving a small flap of skin on the back of the hock joint (see top photo, page 93). This will help prevent the flesh on the drumstick from drawing up and exposing the bone during the roasting process.

Remove the oil sac on the back near the tail as it sometimes gives a peculiar flavor to the meat. This is removed with a wedge-shaped cut (see bottom photo, page 93).

To remove the crop, windpipe, gullet and neck, cut the head off and slit the skin down the back of the neck to a point between the wings, as shown on pages 94-95. Separate the skin from the neck and then from the gullet and windpipe. Follow the gullet to the crop and remove by cutting below the crop. Cut off the neck as close to the shoulders as possible. A pair of pruning shears is handy for this purpose; it can also be done by cutting around the base of the neck with a knife and then breaking and removing with a twisting motion.

When cutting off shanks and feet at the hock, leave a flap of skin behind the hock.

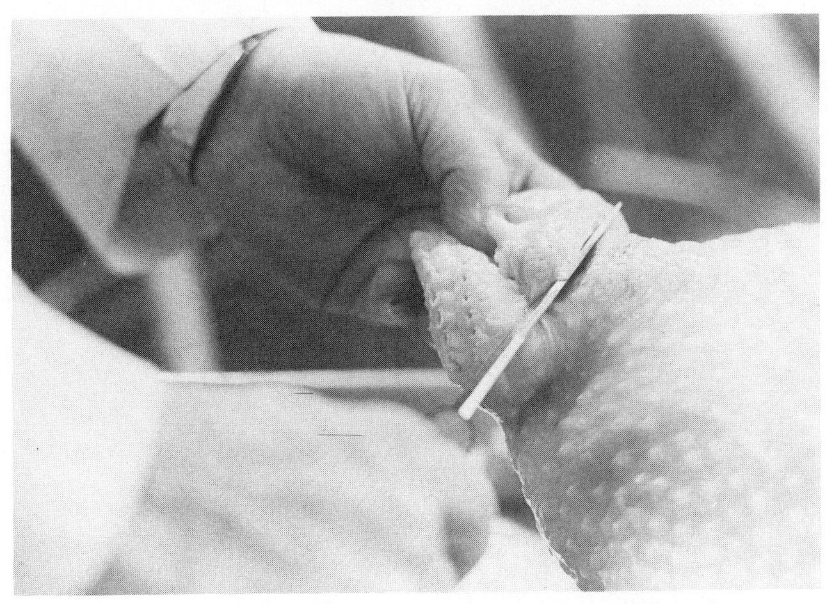

Remove the oil sac near the tail.

To remove the crop, first cut off the head.

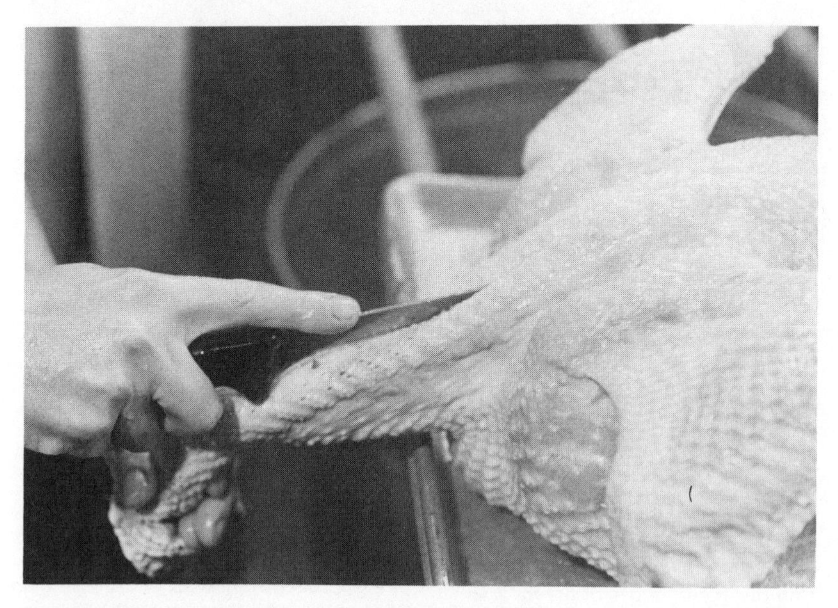

Next slit the skin on the back of the neck.

Remove crop, being careful to cut below it.

Cut off the neck with heavy shears.

Loosen the vent by making a circular cut around it. Do this carefully to avoid cutting into the intestine (see below). Remove the viscera through a short horizontal cut approximately 1½ to 2 inches between the vent and the tip of the keel bone. Make the horizontal cut about 3 inches long. Break the lungs, liver and heart attachments carefully by inserting the hand through the rear opening. Loosen the intestines by working the fingers around them and breaking the tissues that hold them. Remove the viscera through the rear opening in one mass by hooking two fingers over the gizzard, cupping the hand and using a gentle pull and slight twisting motion (see photo, page 97).

Remove the gonads, lungs, and kidneys. The lungs are attached to the ribs on either side of the backbone. These can be removed by using the index finger to break the tissues attaching them to the ribs. Merely insert a finger between the ribs and scrape the lungs loose. The lungs appear pink and spongy. The gonads are also attached to the backbone.

After removing all the organs, wash inside with a hose or under a

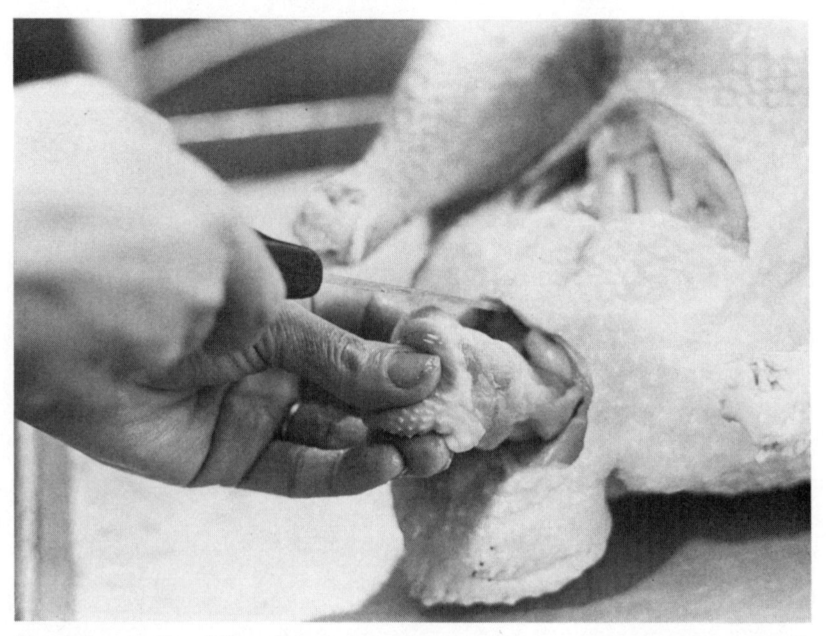

Carefully cut around the vent to loosen it.

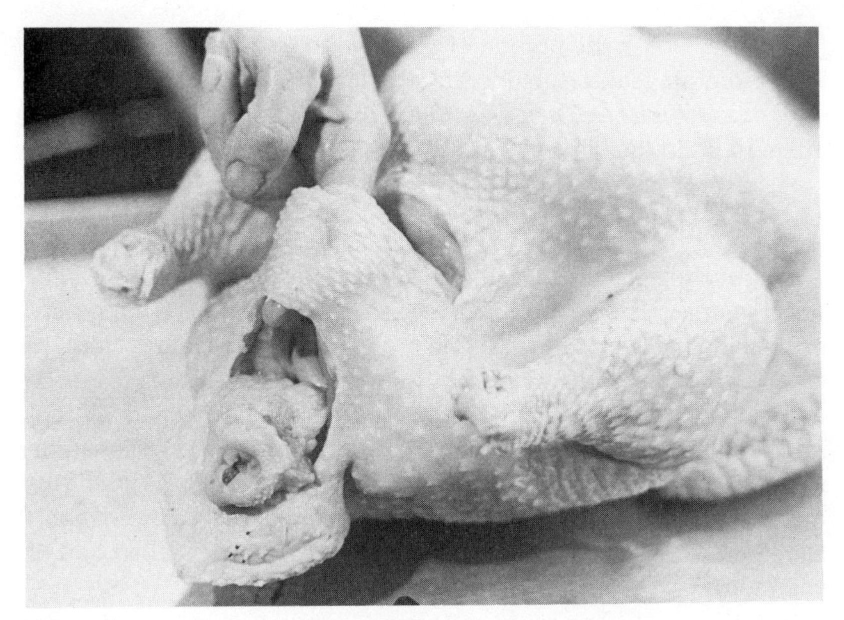

Make a short cut above the vent.

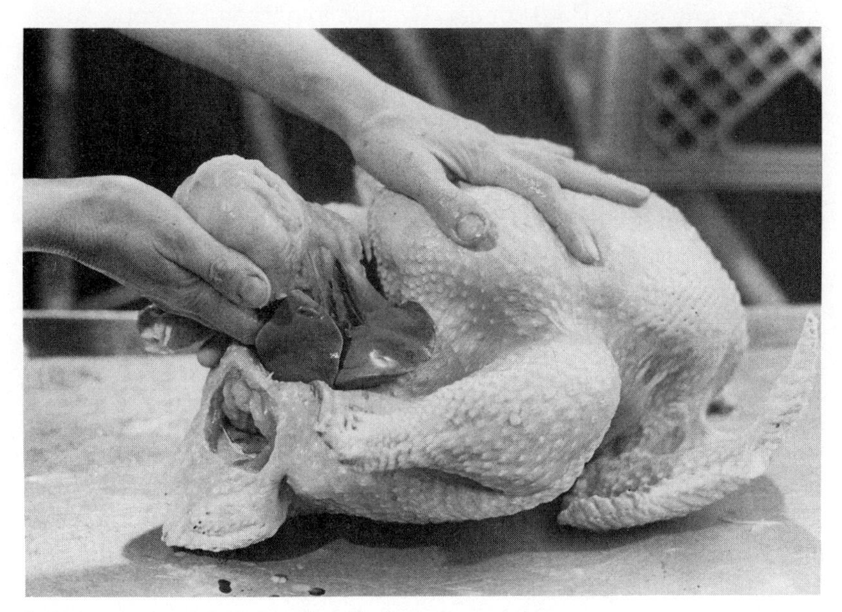

Twisting gently, pull viscera through the cut.

faucet. Also wash the outside of the bird removing all adhering dirt, loose skin, pin feathers, blood or singed hairs. Hang the bird so as to drain the water from the body cavity. Twenty minutes should be adequate to drain the birds thoroughly.

Cleaning the Giblets

Remove the gall bladder, the green sac attached to the liver, without breaking it. If the gall bladder is broken while removing the viscera or cleaning the liver, its bile will likely give a bitter, unpleasant taste to any part it contacts, as well as cause a green discoloration.

If the gizzard is cool, and care is used, it may be cleaned without breaking the inner lining. Cut carefully through the thick muscle until a light streak is observed (see photo, page 99). Do not cut into the inner sac or gizzard lining. The gizzard muscle may then be pulled apart with the thumbs and the sac and its contents removed unbroken, if you're lucky.

Trussing

A properly trussed bird appears neat when packaged. Proper trussing also conserves juices and flavors during roasting.

The simplest trussing method is to place the hock joints under the strip of skin between the vent opening and the cut from which the viscera was removed (see photo, page 100). The neck flap may be drawn back between the shoulders and wing tips folded over the shoulders to hold the skin in place, but usually turkeys are packaged with the wings in their natural position.

Chilling and Packaging

It is important to remove the body heat from the birds as soon as possible after killing. If the cooling is done slowly, bacteria can develop and cause spoilage and undesirable flavors.

If birds are to be air cooled, the air temperature should be from 30° to 35°F. The time required to cool the carcasses depends on the birds' sizes and the temperature of the air. Birds to be air cooled should always be packaged to avoid discoloration.

You can cool poultry with water when it is not possible to cool the

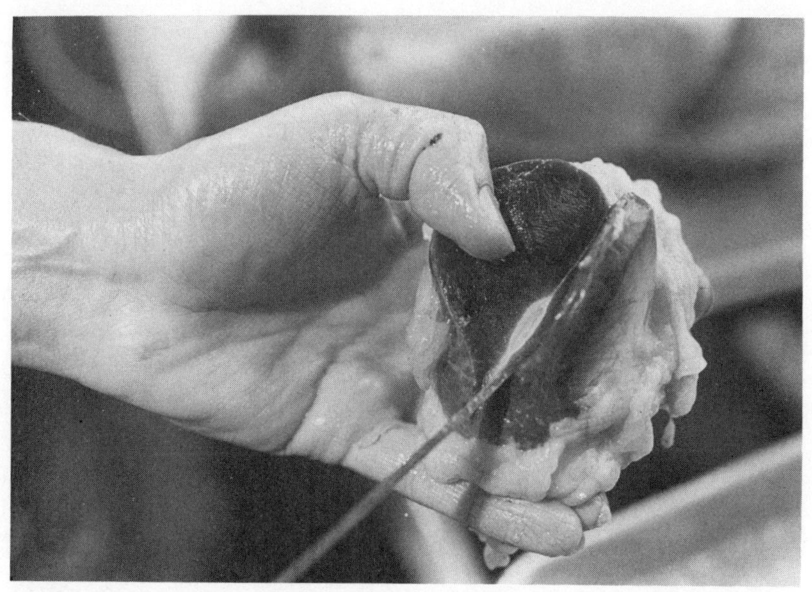

Cut through the gizzard to the light streak.

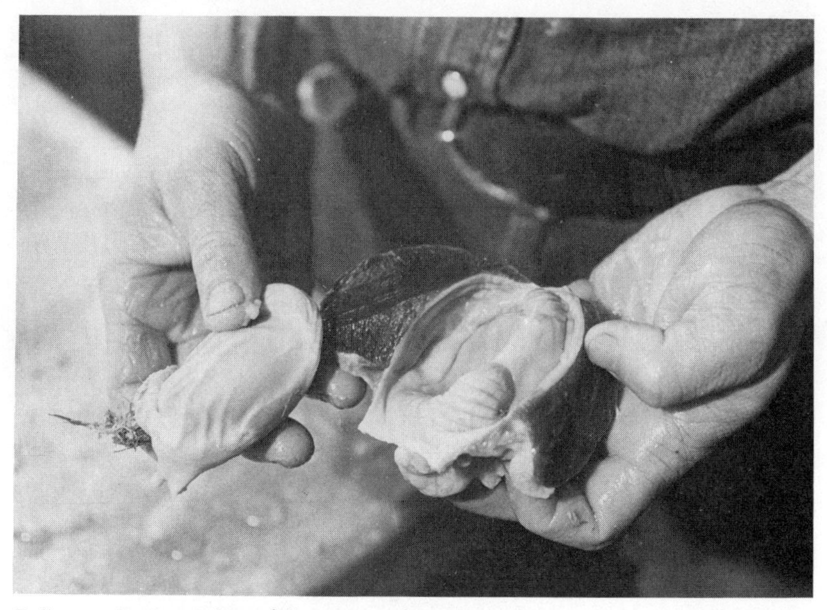

Pull gizzard apart and carefully remove the inner sac.

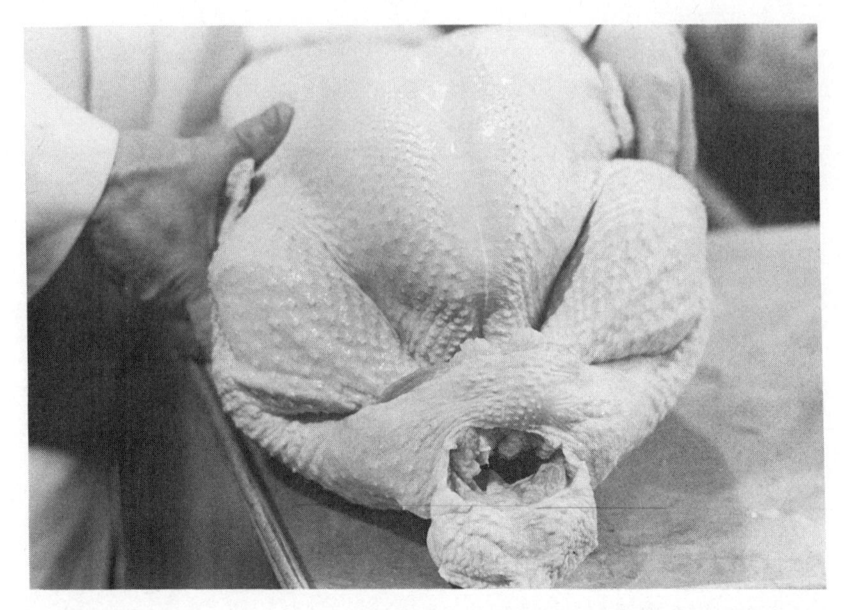

Truss the turkey by tucking the hock joints under the strip of skin above the vent opening.

birds with air. If dressed with excessively high scalding temperatures or for too long a period of time, air-cooled birds may show a blotchy discoloration of the skin. When scalding temperatures are too high, water cooling is the preferred method of cooling the carcasses. Dressed birds may be cooled in tanks of ice water or cold running water. The important factor is to maintain a constant temperature of 34° to 40°F. To cool the bird's internal temperature to 36° to 40° requires five to ten hours in the water, depending again upon the size of the carcass.

Cool and age the birds for approximately eight to ten hours. If eaten or frozen immediately after dressing, the carcasses will tend to be tougher than if aged for a period of time.

Remove the carcasses from the running water, or ice water, and hang them up to dry for 10 to 30 minutes before packaging. Make every effort to get all the water out of the body cavity of the bird before putting it into the bag.

Wrap the giblets, that is the neck, gizzard, heart and liver, in a sheet of waxed paper or a small plastic bag. Giblets can be stuffed into the body cavity or under the neck skin. Wrap the giblets well so that if they should spoil the carcass will not be affected.

There are two types of bags available for poultry. One is the common plastic bag, the other is the so-called *Cryovac* bag. The Cryovac bag shrinks and adheres closely to the bird after placing both in boiling water. It makes a nice-appearing package. It also helps to reduce the amount of water loss during the freezing process. However, good plastic bags are available and will do a satisfactory job of maintaining quality in frozen dressed poultry. The bags should be highly impermeable to moisture to prevent dehydration in freezer storage, which causes toughness.

Birds to be bagged should be trussed thoroughly and then inserted front end first into the plastic bag. After the bird is in the bag, you can remove excess air by applying a vacuum cleaner or by inserting a flexible hose into the top of the bag and then creating a vacuum (see photos below). Merely keep the bag snug around the hose or vacuum cleaner, then suck the air out of the bag. Twist the bag several times and secure it with a wire tie or rubber band.

Fresh-dressed, ready-to-cook turkeys have a shelf life of approx-

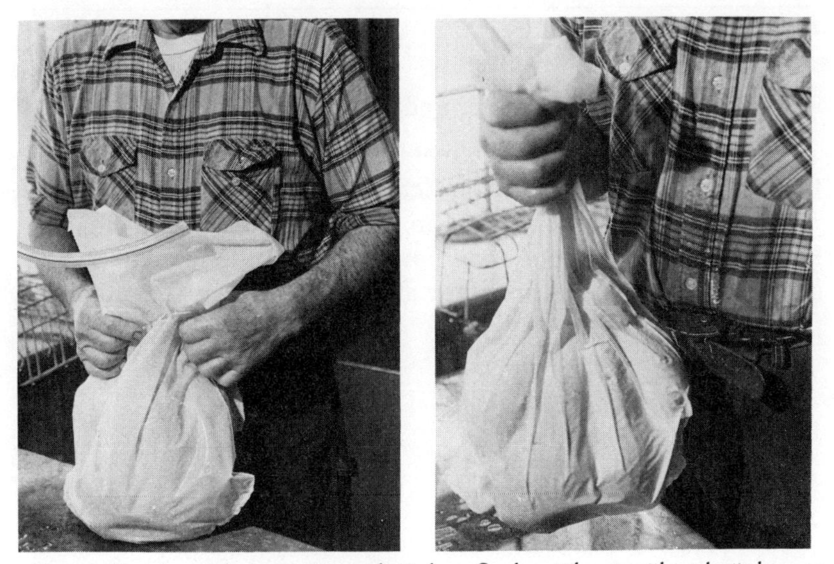

After chilling, insert the carcass in a plastic bag. Suck out the air with a plastic hose or vacuum cleaner and secure the bag with a wire tie.

imately 10 days if refrigerated at a temperature of 29° to 34°F. If you plan to freeze your turkey, do it by the third day after it is dressed and chilled. Chill the poultry to below 40°F. before placing it in the freezer.

The weight loss from live to dressed turkeys varies with the age and the type, as per Table 8.

TABLE 8
DRESSING PERCENTAGES—LIVE TO EVISCERATED WEIGHT

Type of Bird	*Live Weight (Pounds)*	*Blood & Feather Dressed (Percent Loss)*	*Eviscerated (Percent Loss)*
Broilers and fryers	5-6	10	28
Hens — Small	7-9	10	23
Hens — Medium	10-14	9	21
Hens — Large	15-18	8	18
Toms — Small	10-14	10	23
Toms — Medium	15-20	9	21
Toms — Large	24-30	8	18

SOURCE: Penn State University.

State and Federal Grading and Inspection

Some processors of poultry that is sold off the farm are subject to the Poultry Products Inspection Act. There are exemptions for small producers and regulations may vary between states. Check the regulations that apply in your area. For information on grading and inspection programs and how they affect you, contact your state department of agriculture.

CHAPTER 7

Preserving, Cooking and Serving Turkey

In the previous chapter we discussed the slaughter and processing of turkeys in detail—from live to oven ready, packaged in a freezer bag. Turkeys can be held for up to 10 days at about 32°F. If they are to be frozen, they should be frozen by the third or fourth day.

PRESERVING

Freezing

Freezing is probably the simplest and most popular method of preserving the turkey. If stored in a bag that is relatively impervious to moisture, it will maintain its quality for many months. Loss of moisture during freezing and storage will cause a drier, tougher bird when cooked. Usually, oven-ready turkeys are packaged in heavy plastic bags (see pp. 98-102).

Blast freezing (moving air) at temperatures of −20° to −35°F. freezes the carcass quickly and makes for a better-quality frozen product. After the birds are quick frozen, they are stored at −5°F. to −20°F.

Fresh-frozen turkeys are very comparable to fresh turkeys in quality. This is particularly true if directions for proper handling and thawing are followed. More on this later.

The author wishes to acknowledge the assistance of the National Turkey Federation, Reston, Virginia.

Cured, Smoked Turkeys

Turkey that has been cured and smoked is a delicacy. The product has been on the market for many years, particularly during the holiday season. Some stores and shops carry it year round as a specialty item.

The curing and smoking process is a relatively simple one and can be done without expensive ingredients or equipment.

Several different curing preparations are available. Sometimes these cures contain high levels of salt to provide good preservation but these may succeed in masking the delicate poultry flavor.

A salt brine can be prepared at home. For 50 pounds of turkey use the following ingredients:

Water	10 gallons
Salt (non-iodized)	8 pounds (12½ cups)
Brown sugar	3 pounds (7½ cups packed)
Saltpeter	1 cup

If your water is chlorinated, boil it for 15 minutes to get out the chlorine which may affect the curing reaction. Salt brines are corrosive to many metal containers so use a large jar, crock, wooden barrel, stainless steel or plastic tub. Just add the ingredients to the water and stir until dissolved. The brine temperature should be 40°F.

Curing time may be hastened if the turkeys are injected with the brine solution. A syringe with a No. 12 needle works well. Inject deeply into the thickest part of the breast, thighs, drumsticks and joints. Inject an amount equal to approximately 10 percent of the weight of the turkey. Immerse the turkeys in the brine for three days. If the carcasses are not injected with the brine, allow eight to ten days for curing in the brine. Move the turkeys around in the brine at least once during the curing process to make sure the brine penetrates all portions of the carcasses.

Remove the carcasses from the brine, rinse in cold water, drain and wrap in stockinettes or cheesecloth before putting in the smoker.

Hang the birds in a smoke house or other smoker. A good flavor is obtained by using hickory, maple, apple chips or sawdust. The chips are kept moist to prevent flare-ups and to keep up the humidity in the smoker to prevent carcass dehydration.

A ready-to-eat smoked product requires temperatures of 170° to 185°F. for 16 to 24 hours depending on the carcass size. The meat is cooked when the internal temperature of the thickest part of the breast muscle reaches 160°F.

A smoking temperature of approximately 120°F. for 24 hours makes for an excellent uncooked product. To cook cold, uncooked, smoked turkey, use an oven temperature of 275°F. and roast till done (internal temperature 165°F.).

Another recipe for curing and smoking smaller amounts of turkey has been developed by the National Turkey Federation. The accompanying Table 9 provides different brine solutions suitable for different-sized birds.

To prepare for curing and smoking according to the Federation's methods, first place turkey in plastic container, measure water and pour over turkey until water covers about two inches over turkey. Remove turkey. Measure the amount of water needed to cover the turkey and add salt, sugar and salt peter to water, using proportions according to amount of water used (see Table 9). Grind spices in blender and add to brine, stirring vigorously until salt, sugar and salt peter are dissolved. Place turkey in brine and allow it to cure for 2 to 4 days in refrigerator. Remove from brine, dry with paper towels before placing on rack or rotisserie over hot coals.

TABLE 9

BRINE SOLUTION (APPROXIMATELY 10%)

Water	1 gallon	2 gallons	3 gallons	4 gallons	5 gallons	6 gallons	7 gallons
Salt	1 cup	2½ cups	4 cups	5½ cups	7 cups	8½ cups	10 cups
Sugar	⅓ cup	⅔ cup	1 cup	1⅓ cups	1⅔ cups	2 cups	2⅔ cups
Saltpeter	2⅔ tsp.	5⅓ tsp.	8 tsp.	10⅔ tsp.	13⅓ tsp.	16 tsp.	18⅔ tsp.
Bay Leaves	3	6	9	12	15	18	21
Coriander Seeds	3	6	9	12	15	18	21
Whole Cloves	4	8	12	16	20	24	28
Whole Peppercorns	8	16	24	32	40	48	56

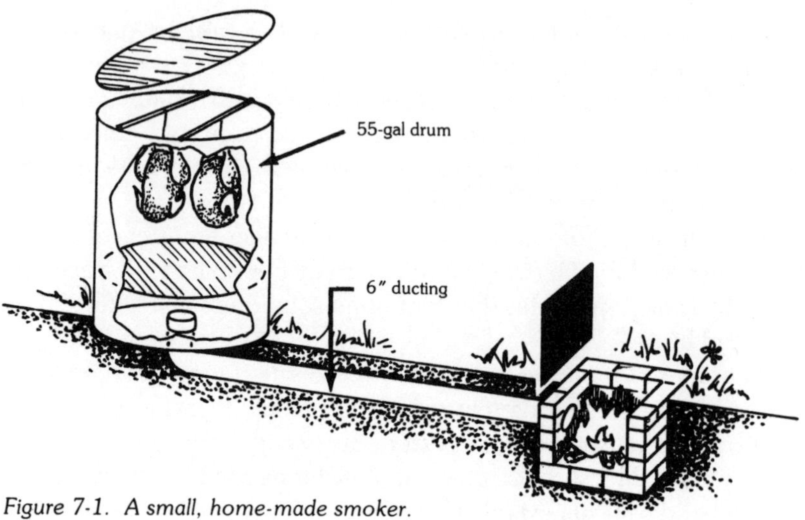

Figure 7-1. A small, home-made smoker.

For smoking, add a few soaked hickory or fruitwood chips every half hour. Keep grill or rotisserie covered to keep smoke in; add more charcoal as needed. Meat thermometer should register 160°F. when cooked and ready to eat.

This method of curing and smoking is not a preservative and turkey must be refrigerated.

Several types of smoke houses or smokers are available commercially or can be constructed at home. A plan for a small smoker is shown in Figure 7-1. If you need something more elaborate and permanent, write to the Government Printing Office, Washington, D.C., and ask for a copy of Farmers' Bulletin No. 2138 published by the U.S. Department of Agriculture.

ROASTING

Preparation

If the turkey is frozen, leave it in the original bag and use one of the following methods to thaw it.

- To thaw slowly, place the turkey on a tray in the refrigerator for 3 to 4 days (24 hours per 5 pounds of turkey).

- If you are in a hurry, immerse the turkey in its water-tight wrapper in cold water and change the water occasionally. This method requires about ½ hour per pound of turkey.

Refrigerate or cook the turkey as soon as it is thawed. If to be stuffed, do it just before cooking. Refreezing an uncooked turkey is not recommended.

After thawing, remove the plastic bag and remove the neck and giblets from the neck and body cavities. Rinse the turkey and wipe dry. The neck and giblets can be cooked for broth for flavoring dressing and for giblet gravy.

If the turkey is to be stuffed, it should be stuffed loosely. Allow ¾ cup of stuffing per pound of oven-ready weight. Unstuffed turkeys require about ½ hour less cooking time. If not stuffed, rub salt in the cavities and, if desired, put in a few pieces of celery, carrots, onion and parsley for flavor.

Tie the legs down or tuck them in the skin flap. The neck skin can be skewered to the back and fold the wing tips back under, in toward the body.

Roasting

Roast the turkey, breast up, on a rack in a shallow pan. Brush the carcass with butter, margarine, or cooking oil if desired. If a meat thermometer is used, insert it into the thickest part of the thigh (not touching the bone).

Roast in an oven set at 325°F. The roasting time chart is a guide to the length of time required (see Table 10). When the thermometer registers 180° to 185°F., the bird is done. (Some cook books suggest 190°F., but there are studies that claim that this temperature results in a decrease in juiciness. On the other hand, at 190°F., odor, flavor, mealiness and doneness increase.)

Foil placed loosely over the turkey will eliminate the need for basting. It may be basted if preferred. Foil should be removed the last half hour to permit a good brown. When the stuffing reaches a temperature of 165°F., it is cooked well. With large birds, it may be difficult to reach this temperature; so, if the bird weighs 24 pounds or more, cooking the stuffing separately is recommended.

The meat thermometer is the most accurate means of determining when the turkey is done. Some people say turkey is done if the drumstick feels soft when pressed with the thumb and forefinger or the drumstick moves easily.

TABLE 10
APPROXIMATE ROASTING TIME FOR TURKEY IN PREHEATED 325°F. OVEN*

Ready-to-Cook Weight	*Approximate Cooking Time*
6 pounds	3 to 3½ hours
8 pounds	3½ to 4 hours
12 pounds	4½ to 5 hours
16 pounds	5½ to 6 hours
20 pounds	6¼ to 6¾ hours
24 pounds	7 to 7½ hours

(½ hour less for unstuffed turkeys)

*Any turkey roasting guide can only give approximate times for several reasons:

1. Turkeys may vary in individual conformation.
2. Exact degree of thawing is difficult to determine.
3. Individual variations in oven temperatures. A domestic oven set at 325 degrees F. can ran range from 300 degrees to 350 degrees.
4. Size of the bird and whether it is stuffed or not.

RECIPES

GOOD GRAVY!

When the turkey is done, pour drippings from roasting pan into a bowl, being sure to leave all the brown particles in the pan. Let fat rise to top of drippings and skim it off into measuring cup. Measure back into roasting pan the amount of fat needed for gravy. (Meat juice under fat should be used as part of liquid.) Set roasting pan with fat over low heat. Blend in flour and cook until bubbly, stirring constantly. Brown fat and flour mixture slightly, if desired. Remove pan from heat, add liquid gradually, stirring until smooth. Return to heat and cook, stirring until mixture is thick. Be sure to scrape brown particles from bottom of pan while cooking. Simmer gently a few minutes, season to taste, and serve hot.

	2 cups gravy (8 servings)	*4 cups gravy (16 servings)*	*6 cups gravy (24 servings)*
Fat	4 Tbsp.	½ cup	¾ cup
Flour	4 Tbsp.	½ cup	¾ cup
Liquid—broth, milk or water	2 cups	4 cups	6 cups

OLD-FASHIONED BREAD STUFFING

1 qt. (4 cups) diced celery
1 cup finely chopped onion
1 cup butter
4 qts. bread cubes, firmly packed (bread, two to four days old)
1 tbsp. salt
2 tsp. poultry seasoning
½ tsp. pepper
1½ to 2 cups broth, milk or water

Cook celery and onion in butter over low heat, stirring occasionally, until onion is tender, but not brown. Meanwhile, blend bread cubes and seasonings. Add celery, onion and butter; toss lightly to blend. Pour the broth, milk or water gradually over surface of bread mixture, tossing lightly. Add more seasoning, as desired. (To increase flavor, try reducing poultry seasoning and adding one teaspoon of sage, marjoram, and thyme and one tablespoon of parsley.) Makes enough stuffing for neck and body cavities of a 14-to 18-pound, ready-to-cook turkey. Note: extra stuffing may be baked in loaf pan or casserole the last hour of turkey cooking. If desired, baste with pan drippings.

CORN BREAD STUFFING

1 pan corn bread, crumbled
1 pkg. dry herbed stuffing mix (or bread crumbs and herbs) for 7-lb. bird
2 cups chopped celery
1 cup chopped onion
½ cup butter
1 cup chicken stock
2 eggs, beaten

Combine corn bread and stuffing mix. Cook celery and onion in butter until tender, but not brown. Add to bread mixture. Add stock and eggs, tossing lightly to blend well. (Add more stock for more moist dressing, if desired.) Stuff lightly into neck and body cavities of turkey; truss. Roast according to standard directions. Extra stuffing may be baked in a 1½-quart casserole, covered, at 325°F. for about 1 hour. Makes enough stuffing for an 18- to 20-pound turkey.

CRANBERRY SAUSAGE DRESSING

2 cups fresh cranberries
1 cup orange juice
⅓ cup sugar
1 pkg. (8 oz.) corn bread stuffing mix
1 lb. fresh pork sausage
½ tsp. baking powder
1 cup finely chopped celery
½ cup finely chopped onion
1 egg
2 tbsp. water

In a saucepan, combine cranberries, orange juice and sugar. Bring to boil, stirring until sugar dissolves. Boil for 5 minutes; cool. Place stuffing mix in large bowl. Break sausage into small pieces over stuffing mix; sprinkle with baking powder. Add celery, onion and egg beaten with water. Toss gently to mix well. Fold in cooled, cooked cranberries. Spoon mixture into 2-quart casserole. Cover tightly and bake at 325°F. for 40 minutes. Remove cover, continue baking 15 minutes longer. Or use to stuff an 8- to 10-pound turkey. Makes 6 to 8 servings.

CARVING A TURKEY

Plan to have the turkey roasted 15 to 30 minutes before it is time to carve it to allow the juices to be absorbed.

SERVING INFORMATION

Ready-to-Cook Weight	*Number of Servings*
4 to 8 pounds	4 to 10
8 to 12 pounds	10 to 20
12 to 16 pounds	20 to 30
16 to 20 pounds	30 to 40
20 to 24 pounds	40 to 50

USES FOR TURKEY PARTS

On occasion, birds may be injured, have leg problems or other conditions which make it advisable to cull them from the flock and salvage usable parts.

When dressing birds, you may find skin lesions, parts that have severe bruises, large breast blisters or other blemishes that detract from the appearance of the dressed carcass. If the birds are otherwise healthy, they can be cut up into parts that can be used in many ways depending upon the age of the birds and the individual's preference.

Turkeys can be cut easily into parts and used in a variety of ways such as barbecuing, pan frying, oven frying or grinding the meat to serve as turkeyburgers.

BARBECUING

Broiler-fryers (usually under 16 weeks of age and weighing around 4 pounds) are excellent for barbecuing. Turkey parts are barbecued in the same way as chicken broilers. A charcoal fire is excellent. Perhaps a gas-fired charcoal oven cooks equally well. Turn the turkey frequently and baste it with a good sauce each time it's turned. Cook until meat thermometer tells you it is done—175° to 180°F. A good sauce recipe consists of the following ingredients: butter or oleo, water, salt and vinegar. To cook approximately 25 pounds of turkey requires the following quantities of sauce materials:

1 lb. of butter or oleo
1 qt. of water
1 qt. of vinegar (cider)
4 tbsp. of salt

Heat the sauce materials to boiling and the sauce is ready to use. It can best be applied with a plastic sprinkling device. You can also apply with a rag tied to the end of a stick.

TURKEYBURGER

Turkey meat can also be ground and used in making turkeyburgers. One interesting recipe for turkeyburgers calls for the following ingredients:

1 lb. of ground turkey
8 single soda crackers (crushed)
2 tbsp. catsup
1 tbsp. lemon juice
1 tbsp. onion flakes
1 tsp. Worcestershire sauce
½ tsp. paprika
4 slices bacon

The ground turkey is combined with the other ingredients and shaped into patties. Wrap the bacon slices around the outer edge and secure with toothpicks. Broil for five or six minutes on each side or until completely done in the center.

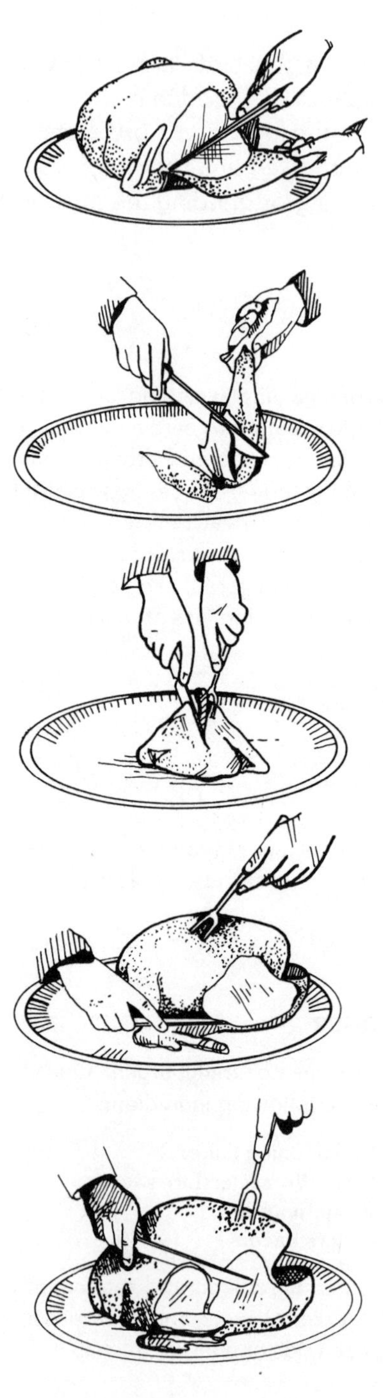

Method 1
(Traditional Method)

1. Remove drumstick and thigh. To remove drumstick and thigh, press leg away from body. Joint connecting leg to the hip will oftentimes snap free or may be severed easily with knife point. Cut dark meat completely from body by following body contour carefully with knife.

2. Slicing dark meat. Place drumstick and thigh on cutting surface and cut through connecting joint. Both pieces may be individually sliced. Tilt drumstick to convenient angle, slicing towards table as shown in illustration.

3. Slicing thigh. To slice thigh meat, hold firmly on cutting surface with fork. Cut even slices parallel to the bone.

4. Preparing breast. In preparing breast for easy slicing, place knife parallel and as close to wing as possible. Make deep cut into breast, cutting right to bone. This is your base cut. All breast slices will stop at this vertical cut.

5. Carving breasts. After making base cut, carve downward, ending at base cut. Start each new slice slightly higher up on breast. Keep slices thin and even.

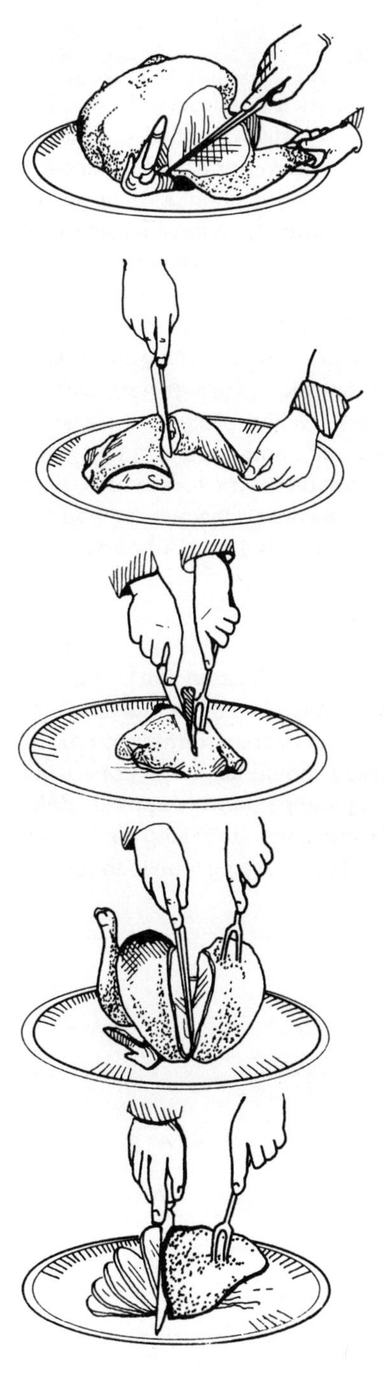

Method 2
(Kitchen-Carving Method)

1. Remove drumstick and thigh by pressing leg away from body. Joint connecting leg to backbone will often snap free or may be severed easily with knife point. Cut dark meat completely from body by following body contour carefully with a knife.

2. Place drumsticks and thigh on separate plate and cut through connecting joint. Both pieces may be individually sliced. Tilt drumstick to convenient angle, slicing towards plate.

3. To slice thigh meat, hold firmly on plate with fork. Cut even slices parallel to the bone.

4. Remove half of the breast at a time by cutting along keel bone and rib cage with sharp knife.

5. Place half breast on cutting surface and slice evenly against the grain of the meat. Repeat with second half breast when additional slices are needed.

PAN FRYING

Birds from 4 to 9 pounds can be pan fried or oven fried successfully. Cut the small whole turkey into pieces to yield 2 drumsticks, 2 thighs, 4 breast pieces, 2 wings, 3 back pieces or pieces as desired. For each 5 pounds of cut-up turkey, blend together $\frac{3}{4}$ cup flour, 1 teaspoon paprika, $\frac{1}{2}$ teaspoon oregano, 2 teaspoons salt, and $\frac{1}{4}$ teaspoon pepper in a bag. To coat with the flour mixture evenly, shake the turkey (2 or 3 pieces at a time) in the bag. Remove turkey from bag.

Save any leftover flour for making gravy. Heat $\frac{1}{2}$ of an inch of oil or fat in heavy skillet until a drop of water just sizzles. Start browning the meaty pieces first, then slip less meaty pieces in between. Turn as necessary to brown and cook evenly (about 20 minutes). When pieces are browned nicely, reduce heat, add 2 tablespoons water, and cover tightly. Cook slowly for 45 to 60 minutes or until the thickest pieces are fork tender. Turn pieces several times for even cooking and browning. Uncover the pan the last 10 minutes to recrisp skin. Total cooking time is 1 to $1\frac{1}{4}$ hours.

OVEN FRYING

To oven fry, cut the turkey into parts and coat with flour as in pan frying. Use a shallow baking pan in a 350°F. oven, melt 1 cup of butter or margarine for each 5 pounds of turkey. Place coated pieces in the pan, turning to coat all sides, then leave skin side down. The turkey should fill the pan one layer deep without crowding and without leaving any pan areas exposed. Bake for 45 minutes. Turn the pieces skin side up and continue baking for another 45 minutes or until the meat is fork tender. Total cooking time is about $1\frac{1}{2}$ hours.

Weights and Measures

3 teaspoons = 1 tablespoon
2 tablespoons = 1 fluid ounce = 6 teaspoons
4 tablespoons = 12 teaspoons = 1/4 cup = 2 fluid ounces
1 cup = 16 tablespoons = 8 fluid ounces
2 cups = 32 tablespoons = 1 pint = 16 fluid ounces
2 pints = 64 tablespoons = 1 quart = 4 level cups
4 quarts = 8 pints = 1 gallon = 16 level cups
16 ounces = 1 pound
6 tablespoons (level) = approximately 1 ounce of dry weight (for wettable powder only)

1 kilogram (kg) = 1000 grams (g) = 2.2. pounds
1 gram (g) = 1000 milligrams (mg) = 0.35 ounce
1 liter = 1000 milliliters (ml) or cubic centimeters (cc) = 1.058 quarts
1 milliliter or cubic centimeter = 0.034 fluid ounces
1 milliliter or cubic centimeter of water weighs 1 gram
1 liter of water weighs 1 kilogram

1 pound = 453.6 grams
1 ounce = 28.35 grams
1 pint of water weighs approximately 1 pound
1 gallon of water weighs approximately 8.34 pounds
1 gallon = 4 quarts = 3.785 liters
1 quart = 2 pints = 0.946 liters
1 pint = 0.473 liters
1 fluid ounce = 29.6 milliliters or cubic centimeters

1 part per million (ppm) = 1 milligram/liter
= 1 milligram/kilogram
= 0.0001 percent
= 0.013 ounces in 100 gallons of water

1 percent = 10,000 ppm
= 10 grams per liter
= 10 grams per kilogram
= 1.33 ounces by weight per gallon of water
= 8.34 pounds/100 gallons of water

0.1 percent	= 1000 ppm	=	1000 milligrams/liter
0.01 percent	= 100 ppm	=	100 milligrams/liter
0.001 percent	= 10 ppm	=	10 milligrams/liter
0.0001 percent	= 1 ppm	=	1 milligram/liter

CUBIC MEASURE

cubic foot	= 1728 cubic inches	cubic yard	= 202 gallons
cubic foot	= 7.48 gallons	cubic yard	= 21.71 bushels
cubic foot	= 0.037 yard	gallon	= 269 cubic inches (dry)
cubic foot	= 0.804 bushels	gallon	= 231 cubic inches (liquid)
cubic yard	= 27 cubic feet	gallon	= 0.134 cubic foot
cubic yard	= 46,656 cubic inches	pound water	= 27.68 cubic inches

pound water = 0.016 cubic foot

Glossary

AIR CELL Air space in the egg usually in the large end.

BEAK Upper and lower mandibles of chickens, peafowl, pheasants, turkeys, etc.

BREAST The foreward part of the body between the neck and keel bone.

BREAST BLISTER Enlarged discolored area or sore in the area of the keel bone.

BROODER Heat source for starting young birds.

BROODINESS Tendency toward the maternal instinct causing the female to set or want to hatch eggs.

CANDLE To determine interior condition of the egg through the use of a special light in a dark room.

CANNIBALISM In the poultry industry, this term refers to the habit of one bird's picking another to the point of injury or death.

CARUNCLES The fleshy, non-feathered area on the neck.

CLOACA The common chamber or receptacle for the digestive, urinary and reproductive tracts.

COCCIDIOSTAT A drug used to control or prevent Coccidiosis.

CROP An enlargement of the gullet where food is stored and readied for digestion.

CULL A bird not suitable to be kept as a breeder or market bird.

CULLING The act of removing unsuitable birds from the flock.

DEBEAK To remove a part of the beak to prevent feather pulling or cannibalism.

EMBRYO A young organism in the early stages of development, as before hatching from the egg.

FLIGHT FEATHERS The large primary and secondary feathers of the wings.

FOOT CANDLE The amount of light striking each and every point on a segment of the inside of a sphere or on a surface area of one square foot all parts of which are one foot from an international candle (a candle of a specified size that emits a specified amount of light). In other words, a foot candle is a measurement of the intensity of a light.

GIZZARD Muscular stomach; its main function is grinding food and partial digestion of proteins.

GULLET OR ESOPHAGUS The tubular structure leading from the mouth to the glandular stomach.

HEN The female turkey.

HOCK The joint of the leg between the lower thigh and the shank.

HOVER Canopy used for brooder stoves to hold the heat down at bird level.

KEEL BONE Breast bone or sternum.

LITTER Soft, absorbent material used to cover floors of poultry houses.

MANDIBLE The upper or lower bony portion of the beak.

MOLT To shed old feathers, which are replaced by new ones.

OIL SAC OR UROPYGIAL GLAND Large oil gland on the back at the base of the tail used by the bird to preen or condition its feathers.

OVA The yolks of eggs.

OVIDUCT Long glandular tube where egg formation takes place, leading from the ovary to the cloaca; it is made up of the funnel, magnum, isthmus, uterus and vagina.

PENDULOUS CROP Crop that is impacted and enlarged and hangs down in an abnormal manner.

PLUMAGE The feathers making up the outer covering of birds.

POULT A young turkey.

POULTRY A term designating those species of birds used by man for food or fiber and can be reproduced under his care, the term includes chickens, turkeys, ducks, geese, pheasants, pigeons and many others.

PRIMARIES The long stiff flight feathers at the outer tip of the wing.

RELATIVE HUMIDITY The percentage of moisture saturation of the air.

ROOST A perch on which birds rest or sleep.

SECONDARIES The large stiff wing feathers adjacent to the body, visible when the wing is folded or extended.

SHANK The scaly portion of the leg below the hock joint and between the thigh and the foot.

SHELL MEMBRANES The two membranes attached to the inner egg shell. They normally separate at the large end of the egg to form an air cell.

SNOOD The fleshy appendage on the head of the turkey.

SPERM OR SPERMATOZOA The male reproductive cells capable of fertilizing the ova.

SPUR The stiff, horny structure on the legs of some birds; found on the inner side of the shanks.

STRAIN Fowl of any breed usually with the breeder's name and which was reproduced by closed flock breeding for five generations or more.

TESTES The male sex glands.

TOM The male turkey.

TRACHEA OR WINDPIPE That part of the respiratory system that conveys air from the larynx to the bronchi and to the lungs and air sacs.

UTERUS The portion of the oviduct where the thin albumen, the shell and shell pigment are added during egg formation.

VAGINA The section of the oviduct that holds the formed egg until it is laid.

VARIETY A subdivision of breed usually distinguished by either color or color and pattern.

VENT OR ANUS The external opening of the cloaca.

YOLK Ovum, the yellow portion of the egg.

References

Aho, William A. and Talmadge, Daniel W. *Incubation and Embryology of the Chick*. Storrs, Ct.: Cooperative Extension Service, College of Agriculture and Natural Resources, University of Connecticut.

Jensen, Leo. "Growth Rate and Feed Consumption Standards." *Turkey World*, Mount Morris, Ill.: January-February 1981.

Jordan, H.C. and Schwartz, L.D. *Home Processing of Poultry*. University Park, Pa.: Penn State University, College of Agriculture, Extension Service.

Jordan, H.C. "Production of Market Turkeys." Correspondence Courses in Agriculture, Course 106, Lesson 1, University Park, Pa.: Penn State University Extension Service.

Marsden, Stanley J. *Turkey Production*. Agricultural Handbook No. 393. Washington, D.C.: Agricultural Research Service, USDA.

Mercia, L.S. *Killing, Picking and Cooling Poultry*. 4-H publication. Burlington, Vt.: The Vermont Extension Service, University of Vermont.

Mercia, L.S. *Raising Poultry the Modern Way*. Pownal, Vt. : Garden Way Publishing Co.

Mercia, L.S. *The Small Turkey Flock*. Burlington, Vt.: The Vermont Extension Service, University of Vermont.

Moyer, D.D. *Virginia Turkey Management*. Publication 302. Blacksburg, Va.: Extension Division, Virginia Polytechnic Institute.

Extension Poultrymen in New England. *Poultry Management and Business Analysis Manual for the 80's*. The New England Cooperative Extension Services.

Extension Poultry Specialists

ALABAMA

Auburn Extension Center
Country Courthouse
Box 1904
Decatur, AL 35601

Auburn University
Poultry Science Department
Auburn, AL 36849

ARIZONA

University of Arizona
Agricultural Sciences Building
Tucson, AZ 85721

ARKANSAS

University of Arkansas
Box 391
Little Rock, AR 72203

University of Arkansas
Department of Animal Sciences
R-C123
Fayetteville, AR 72701

CALIFORNIA

University of California
Kearney Agricultural Center
9240 South Riverbend Avenue
Parlier, CA 93648

University of California
777 East Rialto Avenue
San Bernardino, CA 92415-0730

University of California
733 County Center III Court
Modesto, CA 95355

University of California
Dept. of Avian Science
Davis, CA 95616-8532

University of California
Cooperative Extension Bldg.
Highland Hall
Riverside, CA 92521

COLORADO

University of Colorado
Fort Collins, CO 80523

CONNECTICUT

University of Connecticut
Dept. of Animal Science U-40
3636 Horsebarn Road Ext.
Storrs, CT 06269-4040

University of Connecticut
Cooperative Extension System
562 New London Turnpike
Norwich, CT 06360

DELAWARE

University of Delaware
Research and Education Center
RD 2, Box 48
Georgetown, DE 19947

FLORIDA

University of Florida
Poultry Science Dept.
Gainesville, FL 32611

GEORGIA

University of Georgia
Dept. of Poultry Science
Athens, GA 30602

HAWAII

University of Hawaii
1800 East-West Road
Honolulu, HI 96844

IDAHO

University of Idaho
Route 8, Box 8478
Caldwell, ID 83605

ILLINOIS

University of Illinois
Animal Science Dept.
324 Mumford Hall
Urbana, IL 61801

INDIANA

Purdue University
Lafayette, IN 47907

IOWA

Iowa State University
201 Kildee Hall
Ames, IA 50010

KANSAS

Kansas State University
Leland Call Hall
Manhattan, KS 66506

KENTUCKY

University of Kentucky
Agricultural Science Center
Lexington, KY 40506

LOUISIANA

Louisiana State University
Knapp Hall
Baton Rouge, LA 70803

MAINE

University of Maine
Hitchner Hall
Orono, ME 04469

University of Maine
375 Main Street
Rockland, ME 04841

MARYLAND

University of Maryland
Dept. of Poultry Science
College Park, MD 20742

MASSACHUSETTS

University of Massachusetts
Stockbridge Hall, Room 307A
Amherst, MA 01003

University of Massachusetts
Worcester County Extension Service
36 Harvard Street
Worcester, MA 01608

MICHIGAN

Michigan State University
104 Anthony Hall
East Lansing, MI 48823

MINNESOTA

University of Minnesota
St. Paul, MN 55108

MISSISSIPPI

Mississippi State University
Box 5425
Mississippi State, MS 39762

MISSOURI

University of Missouri
Poultry Building T14
Columbia, MO 65201

NEBRASKA

University of Nebraska
Lincoln, NE 68583

NEW HAMPSHIRE

University of New Hampshire
215 Kendall Hall
Durham, NH 03824

NEW JERSEY

Cook College
Rutgers—The State University
Bartlett Hall, Room 106
New Brunswick, NJ 08903-0231

NEW MEXICO

New Mexico State University
Dept. of Animal and
Range Science
Box 31
Las Cruces, NM 88003

NEW YORK

Cornell University
Farm and Home Center
249 Highland Avenue
Rochester, NY 14620

Cornell University
Dept. Poultry Science
Rice Hall
Ithaca, NY 14853

Cooperative Extension Center
1050 West Genessee Street
Syracuse, NY 13204

NORTH CAROLINA

North Carolina State University
Box 5307 Scott Hall
Raleigh, NC 27650

NORTH DAKOTA

North Dakota University
Dept. of Animal Sciences
Fargo, ND 58105

OHIO

Ohio State University
674 West Lane Avenue
Columbus, OH 43210

OKLAHOMA

Oklahoma State University
Stillwater, OK 74074

OREGON

Oregon State University
Poultry Science Dept.
Corvallis, OR 97331

PENNSYLVANIA

Pennsylvania State University
Poultry Science Dept.
University Park, PA 16802

PUERTO RICO

University of Puerto Rico
Animal Industry Dept.
Mayaguez, PR 00708

RHODE ISLAND

University of Rhode Island
Kingston, RI 02881

SOUTH CAROLINA

Clemson University
Box 378
York, SC 29745

Clemson University
132 Poultry and Animal
Science Bldg.
Clemson, SC 29631

SOUTH DAKOTA

South Dakota State University
Animal Science Complex
Brookings, SD 57007

TENNESSEE

University of Tennessee
Box 1071
Knoxville, TN 37901

TEXAS

Texas A&M University
Poultry Science Dept.
College Station, TX 77843

UTAH

Utah State University
Logan, UT 84322

VERMONT

University of Vermont
College of Agriculture
South Burlington, VT 05401

VIRGINIA

Virginia Polytechnic Institute and
State University
Blacksburg, VA 24061

WASHINGTON

Washington State University
Western Washington Research and
Extension Center
Puyallup, WA 98371

Washington State University
College of Agriculture
Pullman, WA 99163

WEST VIRGINIA

West Virginia University
1066 Agricultural Sciences Bldg.
Morgantown, WV 26506

WISCONSIN

University of Wisconsin
Poultry Science Dept.
Animal Science Bldg.
1675 Observatory Drive
Madison, WI 53706

Sources of Supplies and Equipment

(United States and Canada)

Many of the supplies and equipment needed for the small poultry flock may be found at the local feed store, hatchery, or other agricultural supply outlets. Some of the large mail-order houses such as Sears also handle agricultural supplies.

As an aid to locating sources of certain items not readily available in some areas, a partial list of supplies and equipment and possible sources is included below.

Space would not permit a complete listing of products or company names. There is no intent to recommend one source over another.

Bags — Dressed Poultry

Berry Hill Ltd.
St. Thomas, Ontario
Canada N5P1B0

Dow Chemical Co.
Flexible Packaging Sales
James Salvage Building
Midland, MI 48640

W. R. Grace and Co.
Cryovac Division, Box 464
Duncan, SC 29334

Gordon Johnson Industries
2519 Madison Avenue
Kansas City, MO 64108

Mobil Chemical Co.
Plastics Div. Tech Center
Macedon, NY 14502

Val-A Company
700 W. Root Street
Chicago, IL 60609

Bands, Identification
(Leg and wing tags)

Berry Hill Ltd.
St. Thomas, Ontario
Canada, N5P1B0

Gey Band and Tag Co. Inc.
P.O. Box 363
Norristown, PA 19401

National Band & Tag Co.
721 York Street
Newport, KY 41071

Brooders

Beacon Industries, Inc.
100 Railroad Avenue
Westminster, MD 21157

Berry Hill Ltd.
St. Thomas, Ontario
Canada, N5P1B0

Big Dutchman
P.O. Box 9
New Holland, PA 17557

Brower Equipment Co.
P.O. Box 88
Houghton, IA 52631

Cyclone Intenational, Inc.
P.O. Box 1017
Holland, MI 49422

Lyon Electric Co.
2765A Main Street
P.O. Box 3307
Chula Vista, CA 92011

Port Williams Agencies Ltd.
Port Williams, Nova Scotia
Canada, B0B1T0

Shenandoah Mfg. Co. Inc.
P.O. Box 839
Harrisonburg, VA 22801

A.R. Wood — Northco
Luverne, MN 56156

Cannibalism Control Equipment

Berry Hill Ltd.
St. Thomas, Ontario
Canada, N5P1B0

Gey Band and Tag Co., Inc.
P.O. Box 363
Norristown, PA 19401

Kuhl Corp.
P.O. Box 26, Kuhl Road
Flemington, NJ 08822

Lyon Electric Co.
2765A Main Street
P.O. Box 3307
Chula Vista, CA 92011

NASCO
Fort Atkinson, WI 53538

National Band & Tag Co.
721 York Street
Newport, KY 41071

Egg Baskets

Agri-Sales Associates, Inc.
212 Louise Avenue
Nashville, TN 37203

Beacon Industries, Inc.
100 Railroad Avenue
Westminster, MD 21157

Kuhl Corporation
P.O. Box 26, Kuhl Road
Flemington, NJ 08822

Egg Candlers (Hand), Scales, and Washers

Berry Hill Ltd.
St. Thomas, Ontario
Canada, N5P1B0

Beacon Industries, Inc.
100 Railroad Avenue
Westminster, MD 21157

Brower Equipment Co.
P.O. Box 88
Houghton, IA 52631

Kent Company
13145 Coronado Drive
N. Miami, FL 33181

Kuhl Corporation
P.O. Box 26, Kuhl Road
Flemington, NJ 08822

W. Murray Clark Ltd.
Caledonia, Ontario
Canada, N5P1B0

NASCO
Fort Atkinson, WI 53538

National Poultry Equipment Co.
1520 Standiford Avenue
Modesto, CA 95350

Val-A Company
700 W. Root Street
Chicago, IL 60609

Feeders and Waterers

Beacon Industries, Inc.
100 Railroad Avenue
Westminster, MD 21157

Big Dutchman
P.O. Box 9
New Holland, PA 17557

Shenandoah Mfg. Co., Inc.
P.O. Box 839
Harrisonburg, VA 22801

Incubators — Small

Berry Hill Ltd.
St. Thomas, Ontario
Canada, N5P1B0

Brower Equipment
P.O. Box 88
Houghton, IA 52631

Kent Company
13145 Coronado Drive
N. Miami, FL 33181

Kuhl Corporation
P.O. Box 26, Kuhl Road
Flemington, NJ 08822

Lyon Electric Co.
2765A Main Street
P.O. Box 3307
Chula Vista, CA 92011

Petersyme Incubator Co.
Gettysburg, OH 45328

Killing Cones, Knives, and Other Dressing Equipment

Ashley Machine, Inc.
901 Carver Street
P.O. Box 2
Greensburg, IN 47240

Kuhl Corporation
P.O. Box 26, Kuhl Road
Flemington, NJ 08822

NASCO
Fort Atkinson, WI 53538

Nunnally Enterprises, Inc.
4550 North Star Way
Modesto, CA 95356

Nests

Beacon Industries, Inc.
100 Railroad Avenue
Westminster, MD 21157

Big Dutchman
P.O. Box 9
New Holland, PA 17557

Cyclone International, Inc.
P.O. Box 1017
Holland, MI 49422

Shenandoah Manufacturing Co., Inc.
P.O. Box 839
Harrisonburg, VA 22801

Processing Equipment
(Pickers, Scalders, etc.)

Ashley Machine, Inc.
901 Carver Street, P.O. Box 2
Greensburg, IN 47240

Kent Company
13145 Coronado Drive
N. Miami, Florida 33181

Kuhl Corporation
P.O. Box 26, Kuhl Road
Flemington, NJ 08822

Pickwick Company
P.O. Box 756
1120 Glass Road
Cedar Rapids, IA 52406

Smoke Houses

Kent Company
13145 Coronado Drive
N. Miami, FL 33181

Wax — Defeathering

Continental Machine Co.
P.O.Box 2936
222 Myrtle Street, S.E.
Gainesville, GA 30501

Pickwick Co.
P.O. Box 756
1120 Glass Road
Cedar Rapids, IA 52406

Poultry Diagnostic Laboratories

ALABAMA

Alabama Veterinary Diagnostic
Laboratory
P.O. Box 2209
Auburn, AL 36830

Poultry Producers Diagnostic
Laboratory
1724-A Second Avenue NW
Cullman, AL 35055

Alabama Veterinary Diagnostic
Laboratory
192 Boswell Street
Elba, AL 36323

ARIZONA

Department of Veterinary Science
University of Arizona
Tucson, AZ 85721

ARKANSAS

Arkansas Livestock and Poultry
Commission
Diagnostic Laboratory
Highway 71 N. Box 766
Springdale, AR 72764

CALIFORNIA

University of California
Veterinary Diagnostic Laboratory
System (CVDLS):

West Health Sciences Drive
P.O. Box 1770
Davis, CA 95617-1770

2789 South Orange Avenue
Fresno, CA 93725

105 West Central Avenue
San Bernardino, CA 92412

18830 Road 112
Tulare, CA 93274

1550 N. Soderquist
P.O. Box P
Turlock, CA 95380

COLORADO

Veterinary Diagnostic Laboratory
Colorado State University
Fort Collins, CO 80523

CONNECTICUT

Department of Pathobiology
University of Connecticut
Box U-89
Storrs, CT 06269-3089

DELAWARE

Division of Standards and Inspection
State Department of Agriculture,
Poultry and Animal Health Section
P.O. Drawer D
Dover, DE 19901

Diagnostic Laboratory
University of Delaware Substation
Route 2, Box 47
Georgetown, DE 19947

FLORIDA

Bureau of Diagnostic Laboratories
Dade City Branch
P.O. Box 1031
Dade City, FL 33525

Florida Department of Agriculture
Bureau of Diagnostic Laboratories
Kissimmee Branch
P.O. Box 420460
Kissimmee, FL 34742-0460

Bureau of Diagnostic Laboratories
Miami Branch
8701 NW 58th Street
Miami, FL 33178

GEORGIA

Georgia Poultry Laboratory Number 1
P.O. Box 20
Oakwood, GA 30566

Georgia Poultry Laboratory Number 2
P.O. Box 349
Canton, GA 30114

Georgia Poultry Laboratory Number 4
1126 Lamar Street
Dalton, GA 30720

Georgia Poultry Laboratory Number 8
Route 3, Douglas Airport
Douglas, GA 31533

Veterinary Diagnostic and
Investigational Laboratory
College of Veterinary Medicine
P.O. Box 1389
Tifton, GA 31793

HAWAII

Hawaii Department of Agriculture
Veterinary Laboratory Branch
1428 South King Street
Honolulu, HI 96814

ILLINOIS

Animal Disease Laboratory
Shattuc Road
Centralia, IL 62801

INDIANA

Animal Disease Diagnostic
Laboratory S.I.P.A.C.
Poultry Pathologist
RR #1
Dubois, IN 47527

Animal Disease Diagnostic
Laboratory
Purdue University
West LaFayette, IN 47907

IOWA

Veterinary Diagnostic Laboratory
Iowa State University
Ames, IA 50010

KANSAS

Veterinary Diagnostic Laboratory
College of Veterinary Medicine
Veterinary Medical Center
Manhattan, KS 66506

KENTUCKY

Livestock Disease Diagnostic Center
RR 6
Newton Pike
Lexington, KY 40505

LOUISIANA

Central Louisiana Livestock
Diagnostic Laboratory
Route 2, Box 51-F
Lecompte, LA 71346

MAINE

University of Maine
Pathology Diagnostic and Research
Laboratory
Hitchner Hall
Orono, ME 04473

Maine Poultry Consultants
Box 788
Waterville, ME 04901

MARYLAND

Maryland Department of Agriculture
P.O. Box J
Salisbury, MD 21801

MASSACHUSETTS

Paige Laboratory
University of Massachusetts
Amherst, MA 01002

Avian Diagnostic Laboratory
240 Beaver Street
Waltham, MA 02154

MICHIGAN

Animal Health Diagnostic Laboratory
P.O. Box 30076
Lansing, MI 48909

MISSISSIPPI

Mississippi Veterinary Diagnostic
Laboratory
P.O. Box 4389
Jackson, MS 39216

MISSOURI

Ralston Purina Veterinary Laboratory
Checkerboard Square
St. Louis, MO 63188

MONTANA

State of Montana, Animal Health Division
Diagnostic Laboratory
P.O. Box 997
Bozeman, MT 59715

NEBRASKA

Diagnostic Laboratory
Department of Veterinary Science
University of Nebraska
Lincoln, NE 68503

Veterinary Science Laboratory
University of Nebraska
North Platte Station
North Platte, NE 69101

NEVADA

Nevada Department of Agriculture
Animal Disease Laboratory
P.O. Box 11100
Capitol Hill Avenue
Reno, NV 89510

NEW HAMPSHIRE

Veterinary Diagnostic Laboratory
University of New Hampshire
Durham, NH 03824

NEW JERSEY

New Jersey Animal Health Diagnostic Laboratory
John Fitch Plaza, CN 330
Trenton, NJ 08625

NEW MEXICO

New Mexico Veterinary Diagnostic Laboratory
700 Camino De Salud NE
Albuquerque, NM 87106

NEW YORK

Cornell University
Duck Research Laboratory
Box 217 Old Country Road
Eastport, NY 11941

Department of Avian & Aquatic Animal Medicine
College of Veterinary Medicine at Cornell University
Ithaca, NY 14853

NORTH CAROLINA

Rollins Animal Disease Diagnostic Laboratory
P.O. Box 12223
Cameron Village Station
Raleigh, NC 27605

OHIO

Ohio Department of Agriculture
Animal Disease Diagnostic Laboratory
Reynoldsburg, OH 43068

Ohio State University
Department of Veterinary Clinical
Sciences, Clinical Pathology
Laboratory
1935 Coffey Road
Columbus, OH 43210

OKLAHOMA

Oklahoma Animal Disease Diagnostic
Laboratory
College of Veterinary Medicine
Oklahoma State University
Stillwater, OK 74074

OREGON

Oregon State University
Veterinary Diagnostic Laboratory
Corvallis, OR 97331

State Federal Veterinary Laboratory
635 Capitol Street NE
Salem, OR 97310

PENNSYLVANIA

Animal Diagnostic Laboratory
Pennsylvania State University
University Park, PA 16802

Regional Diagnostic Laboratory
5349 William Flynn Highway
Route 8
Gibsonia, PA 15044

Poultry Diagnostic Laboratory
New Bolton Center
Kennett Square, PA 19348

Pennsylvania Department of
Agriculture
Bureau of Animal Industry
Laboratory
Summerdale, PA 17093

Regional Diagnostic Laboratory
R 92 South
Tunkhannock, PA 18657

PUERTO RICO

Puerto Rico Animal Diagnostic
Laboratory
P.O. Box E
Dorado, PR 00646

RHODE ISLAND

Diagnostic Laboratory
Department of Animal Pathology
University of Rhode Island
Kingston, RI 02881

Rhode Island Department of Health
Animal Disease Laboratory
Health Laboratory Building
50 Arms Street
Providence, RI 02904

SOUTH CAROLINA

Clemson University
Livestock-Poultry Health Department
P.O. Box 102406
Columbia, SC 29224-2406

TENNESSEE

C. E. Kord Animal Disease
Laboratory
P.O. Box 40627
Melrose Station
Nashville, TN 37204

TEXAS

Texas A&M University
Poultry Disease Laboratory
College of Veterinary Medicine
College Station, TX 77843

Texas A&M University
Poultry Disease Laboratory
P.O. Box 84
Gonzales, TX 78629

UTAH

Utah State University
Veterinary Diagnostic Laboratory
Logan, UT 84321

VERMONT

Animal Health
University of Vermont
655-C Spear Street
South Burlington, VT 05403

VIRGINIA

Division of Animal Health and
Dairies
Regulatory Laboratory
116 Reservoir Street
Harrisonburg, VA 22801

Division of Animal Health and
Dairies
Regulatory Laboratory
Ivor, VA 23866

Division of Animal Health and
Dairies
Regulatory Laboratory
234 West Shirley Avenue
Warrentown, VA 22186

WASHINGTON

Washington Animal Disease
Diagnostic Laboratory
P.O. Box 2037, College Station
Washington State University
Pullman, WA 99164

Poultry Diagnostic Laboratory
Western Washington Research and
Extension Center
Washington State University
Puyallup, WA 98371

WEST VIRGINIA

State Federal Cooperative Animal
Health Laboratory
Room B 86
Capitol Building
Charleston, WV 25305

West Virginia Regional Animal
Health Laboratory
Box P
Moorefield, WV 26836

WISCONSIN

Regional Animal Health Laboratory
1418 LaSalle Avenue
Barron, WI 54812

Central Animal Health Laboratory
6101 Mineral Point Road
Madison, WI 53705

WYOMING

Wyoming State Veterinary Laboratory
1174 Snowy Range Road
Laramie, WY 82070

Index